Jones and Laughlins

**Report on Cold-Rolled Iron and Steel**

Jones and Laughlins

**Report on Cold-Rolled Iron and Steel**

ISBN/EAN: 9783743419469

Manufactured in Europe, USA, Canada, Australia, Japa

Cover: Foto ©berggeist007 / pixelio.de

Manufactured and distributed by brebook publishing software (www.brebook.com)

Jones and Laughlins

**Report on Cold-Rolled Iron and Steel**

# REPORT ON COLD-ROLLED

# IRON AND STEEL

AS MANUFACTURED BY

# JONES & LAUGHLINS,

## AMERICAN IRON WORKS,

### PITTSBURGH.

------

By ROBERT H. THURSTON, A. M., C. E.,

PROF. OF ENGINEERING, STEVENS INSTITUTE OF TECHNOLOGY; MEMBER OF
AM. SOCIETY OF CIVIL ENGINEERS; INSTITUTE OF MINING ENGINEERS;
SOCIETE DES INGENIEURES CIVILS; VEREIN DEUTSCHE INGENIEURE;
OESTERRICHISCHE INGENIEURE UND ARCHITEKTEN VEREIN;
INSTITUTION OF ENGINEERS AND SHIPBUILDERS OF SCOT-
LAND, &C., &C., &C.; ASSOCIATE BRITISH INSTITUTION
OF NAVAL ARCHITECTS; FELLOW OF NEW YORK
ACADEMY OF SCIENCES; AM. ASSOCIATION FOR
ADVANCEMENT OF SCIENCE,
ETC., ETC., ETC.

PRINTED BY STEVENSON, FOSTER & CO., NO. 48 FIFTH AVENUE.
1878.

# REPORT ON

# COLD-ROLLED IRON.

## CONTENTS.

# INTRODUCTION.

The investigations here reported were undertaken by the undersigned at the request of Messrs. Jones & Laughlins, proprietors of the American Iron Works, Pittsburgh, in the summer of the year 1877, and have been continued almost uninterruptedly to date. They constitute the most complete research upon the properties of any one of the many metals used in engineering construction that has yet been made, so far as the knowledge of the writer extends. They include tests of finished shafting and round iron from $2\frac{2}{10}$ inches in diameter down to $\frac{3}{8}$ inch, both in tension and by transverse strain, and tests in the Autographic Recording Testing Machine of iron cut from each grade and size.

The tests exhibiting the fact that cold-rolling produces a bar of more uniform strength from surface to centre than is made by the common process of hot-rolling, are as important as the results are novel. Later tests which exhibit the fact that the "mild" or "low" steels, so-called, are benefited by the process are, if possible, of greater value than those of iron, since the use of these mild steels —or, more properly, homogeneous irons—seems certain to result in time in the exclusion of puddled iron and steel from all engineering work.

These investigations have been made in the Mechanical Laboratory of the Department of Engineering of the Stevens Institute of Technology, where one-half of each broken test-piece is retained. The record books of the Laboratory also contain the original records from which the figures here given are taken. Both the retained samples and the records can be seen, with the results of an immense number of other tests, by any one who chooses to examine them. In this work I have been greatly assisted by Mr. J. E. Denton, my principal assistant, who has charge of the

Laboratory, by Mr. T. F. Koczly, Recorder, and especially by Mr. Edward A. Uehling, the observer especially detailed to assist me in this work, to whose intelligence, knowledge and skill I am indebted to an extent which I cannot too fully acknowledge. The amount of fine work demanded in making such nice determination, and the immense amount of calculation involved in the working up of results, can only be understood by the very few who have themselves engaged in such research. The accuracy and neatness of the plates accompanying these reports are due to the skill of Mr. F. T. Thurston, C. E.

The investigations here reported are not the first which have been made. In April of the year 1859 the late Chief Engineer, John P. Whipple, U. S. N., an officer of great experience and of high standing professionally and socially, for whose ability the writer can vouch from personal acquaintance, made a series of experiments upon "bright" or "polished" (cold-rolled) plate-iron, comparing it with the "natural" or ordinary hot-rolled plate, with the following results :

PHILADELPHIA, April 16, 1859.

MR. H. D. KING, Philadelphia :

*Dear Sir*—In compliance with your request, I have tested the samples of "*Polished Plate*" Iron manufactured by Messrs. Jones & Laughlins, Pittsburgh, Pa., in comparison with others of the same quality of iron in the natural state.

The results are shown in the following table :

| Nos. | QUALITY OF IRON. | Sectional Area of Sample. sq. in. | Breaking Weight of Sample. lbs. | Breaking Weight in lbs. per sq. in. | Increase of strength in Polished Iron. lbs. |
|---|---|---|---|---|---|
| 1 | Polished Plate | .1824 | 19,125 | 104,852 | |
| 1 | Natural Plate | .424908 | 22,750 | 53,541 | 51,311 |
| 2 | Bright Plate | .17126 | 16,875 | 93,100 | |
| 2 | Natural Plate | .45152 | 27,000 | 59,797 | 33,403 |
| 3 | Bright Plate | .1589 | 13,125 | 82,600 | |
| 3 | Natural Plate | .424908 | 22,750 | 53,541 | 29,059 |
| 4 | Bright Plate | .1844 | 20,750 | 112,527 | |
| 5 | Bright Plate | .1855 | 21,250 | 114,555 | |

I am, very truly, yours,

JOHN P. WHIPPLE,

*Chief Engineer U. S. Navy.*

During the summer of the same year, Mr. Wm. Fairbairn,[*] who was then and up to the time of his death (1874,) the most distinguished living authority on the use of iron and steel in engineering construction, made a series of experiments upon cold-rolled metal, reporting the following results :

### EXPERIMENT 1.

On a bar of Wrought Iron, in the condition in which it is received from the manufacturer, (BLACK.)

Diameter 1.07 inches................ .........................Area 0.85873 sq. inches.

| | Weights laid on in pounds. | Elongation of a length of 10 inches, in inches. | Breaking Weight per square inch. | |
|---|---|---|---|---|
| | | | In Lbs. | In Tons. |
| 1 | 9,186 | ...... | ......... | ......... |
| 2 | 46,426 | 1.30 | ......... | ......... |
| 3 | 50,846 | 2.00 | 58,628 | 26.178 |

Diameter at point of fracture after the experiment, 0.88 in.

### EXPERIMENT 2.

On a bar similar to the preceding, but Rolled Cold.

Diameter 1.00 inches..... .................................. ........Area 0.7854 sq. inches.

| | Weights laid on in pounds. | Elongation of a length of 10 inches, in inches. | Breaking Weight per square inch. | |
|---|---|---|---|---|
| | | | In Lbs. | In Tons. |
| 1 | 32,590 | 0.01 | ......... | ......... |
| 2 | 37,680 | 0.04 | ......... | ......... |
| 3 | 42,670 | ...... | ......... | ......... |
| 4 | 56,110 | 0.07 | ......... | ......... |
| 5 | 57,585 | 0.08 | ......... | ......... |
| 6 | 60,895 | Elongating | Unbroken. | Unbroken. |
| 7 | 64,255 | rapidly. | 81,812 | 36.533 |

At this point the experiment was discontinued.

[*]Sir Wm. Fairbairn, Baronet; (1869) F. R. S.; LL. D.

## EXPERIMENT 3.

On a bar of Iron, Rolled Cold.

Diameter 1.00 inches.............................. ............................Area 0.7854 sq. inches.

|  | Weights laid on in pounds. | Elongation of a length of 10 inches, in inches. | Breaking Weight per square inch. | |
|---|---|---|---|---|
|  |  |  | In Lbs. | In Tons. |
| 1 | 10,750 | ...... | .......... | .......... |
| 2 | 10,150 | ...... | .......... | .......... |
| 3 | 25,870 | ....... | .......... | .......... |
| 4 | 32,590 | None. | .......... | .......... |
| 5 | 49,135 | ...... | .......... | .......... |
| 6 | 52,495 | ...... | ........ | .......... |
| 7 | 62,575 | 0.6 | ......... | .......... |
| 8 | 69,295 | 0.79 | 88,230 | 39.388 |

Diameter after fracture, 0.85.

## EXPERIMENT 4.

On a bar of similar Iron to the preceding, turned in the lathe.

Diameter 1.00 inches...............................................Area 0.7854 sq. inches.

|  | Weights laid on in pounds. | Elongation of a length of 10 inches, in inches. | Breaking Weight per square inch. | |
|---|---|---|---|---|
|  |  |  | In Lbs. | In Tons. |
| 1 | 10,750 | ...... | .......... | .......... |
| 2 | 19,150 | ...... | .......... | .......... |
| 3 | 27,550 | ...... | .......... | .......... |
| 4 | 30,910 | 0.15 | .......... | .......... |
| 5 | 34,270 | 0.27 | ........ | .......... |
| 6 | 37,630 | 0.48 | ......... | .......... |
| 7 | 40,990 | 0.80 | ........ | .......... |
| 8 | 42,670 | ...... | .......... | .......... |
| 9 | 44,350 | 0 90 | .......... | .......... |
| 10 | 47,710 | 2.20 | 60,746 | 27.119 |

Diameter after fracture, 0.80.

GENERAL SUMMARY OF RESULTS.

| Condition of Bar. | Breaking Weight of Bar in lbs. | Breaking Weight per square inch. | | Strength, the untouched bar being unity. |
|---|---|---|---|---|
| | | In Lbs. | In Tons. | |
| 1 Untouched (black) | 50,346 | 58.628 | 26.173 | 1.000 |
| 3 Rolled Cold.......... | 69,295 | 88.230 | 39.388 | 1.505 |
| 4 Turned................. | 47,710 | 60.746 | 27.119 | 1.036 |

☞ In the above summary, it will be observed that the effect of consolidation by the process of Cold-Rolling is to increase the tensile powers of resistance from 26.17 tons per square inch, to 39.38 tons, being in the ratio of 1 : 1.5, one-half increase of strength gained by the new process of Cold-Rolling.

WILLIAM FAIRBAIRN.

*Manchester*, Aug. 5, 1859.

---

MANCHESTER, August 26, 1859.

SIRS:—In conformity with your request, I carefully inspected your machinery for Cold-Rolling and Polishing Bar Iron, and I have no hesitation in bearing my testimony to its efficiency, and the very perfect manner in which the work was accomplished, both as regards the consolidation of the metal, by which its tenacity is increased, and the roundness and straightness of the bars as they left the machine.

A similar process applied to boiler and bridge plates would not only give greatly increased strength, but would secure a smoothness of finish in their manufacture admirably adapted to enhance the value and increase the importance of iron as a material of construction.

I append the results of some of the experiments on Cold-Rolled Iron.

And remain, Sirs, your obedient servant,

WILLIAM FAIRBAIRN.

## EXPERIMENTS ON THE INCREASE OF COHESIVE STRENGTH IN IRON ROLLED COLD.

| Condition of Bar. | Tensile Strength per sq. in. | | Elongation of 10 inches, in inches. | Ratio of Strength, Black Bar being taken at 1,000. |
|---|---|---|---|---|
| | In Lbs. | In Tons. | | |
| Black Bar, from Rolls.............. | 60,746 | 27.119 | 2 20 | 1.037 |
| Bar turned down to 1 in. diam.... | 58,628 | 26.173 | 2.00 | 1.000 |
| Bar rolled cold to 1 in. diam...... | 88,230 | 39.388 | 0.79 | 1.505 |

The first bar was broken in the condition in which it came from the iron manufacturer; the second was a similar bar turned in the lathe, and the third had been subjected to the process of Cold-Rolling.

It is obvious that the effect of the consolidation in the last case was to increase the strength of the bar in the ratio of 10 to 15.

WILLIAM FAIRBAIRN.

*Manchester, England.*

———

A year later (November 23, 1860,) the late Major Wm. Wade, U. S. A., whose experiments and reports on ordnance and ordnance metal have made his name familiar throughout the civilized world, and whose conscientious and extraordinarily careful and accurate methods of work are known to every one who was so fortunate as to have been familiar with the veteran and his work, also made such a series of tests. His results are here transcribed :

' Summary of the average results obtained from numerous experiments with Bar Iron, rolled while HOT, in the usual manner, compared with the results obtained from the same kinds of Iron, rolled and polished while COLD, by Lauth's patent process.

| | Iron rolled while | | Ratio of Increase by Cold Rolling. | Average rate per ct. of increase. |
|---|---|---|---|---|
| | HOT. | COLD. | | |
| **TRANSVERSE.**—Bars supported at both ends; load applied in the middle; distance between the supports, 30 inches. Weight which gives a permanent set of one-tenth of an inch, viz:— 1½ in. square bars | 3,100 | 10,700 | 3.451 | |
| Round bars, 2 in. diameter | 6,200 | 11,100 | 2.134 | 162½ |
| Round bars, 2¼ " | 6,800 | 15,600 | 2.294 | |
| **TORSION.**—Weight which gives a permanent set of one degree, applied at 25 inches from centre of bars. Round bars, 1¾ in. diameter, and 9 in. between the clamps | 760 | 1,725 | 2.300 | 130 |
| **COMPRESSION.**—Weight which gives a depression, and a permanent set of one-hundredth of an inch to columns 1½ in. long and ⅝ in. diameter. Weight which bends and gives a permanent set to columns 8 in. long and ¾ in. diameter, viz:— Puddled iron | 13,000 | 34,000 | 2.615 | 161¼ |
| | 21,000 | 31,000 | 1.476 | |
| Charcoal bloom iron | 20,500 | 37,000 | 1.804 | 64 |
| **TENSION.**—Weight per sq. inch, which caused rods ¾ in. diam. to stretch and take a permanent set, viz: Puddled iron | 37,250 | 68,427 | 1.837 | 95 |
| Charcoal bloom iron | 42,439 | 87,396 | 2.059 | |
| Weight per square inch, at which the same rods broke, viz: Puddled iron | 65,760 | 83,156 | 1.491 | 72 |
| Charcoal bloom iron | 54,927 | 99,293 | 1.950 | |
| **HARDNESS.**—Weight required to produce equal indentations | 5,000 | 7,500 | 1.500 | 50 |

NOTE.—Indentations made by equal weights, in the centre, and near the edges of the fresh cut ends of the bars, were equal; showing that the iron was as hard in the centre of the bars as elsewhere.

Major Wade also determined the specific gravity of the metal, and, although working with extreme care and making repeated determinations, was unable to detect any increase due to cold-rolling—an unexpected result, but one which has been confirmed by the recent investigations of the writer.

Still another set of tests were made at the Twenty-sixth Exhibition of the Franklin Institute, Philadelphia, the committee reporting as follows:

*Extract from Report on the Twenty-sixth Exhibition of American Manufactures, by the Franklin Institute, Philadelphia:*

Patent Cold-Rolled Polished Shafting, Jones & Laughlins, Pittsburgh, Pa. Deposited by H. D. King.

A handsome display of cold-rolled and polished iron of various sizes, intended for shafting, piston rods or other purposes where turned iron is used; also, flat sheets, suitable for spade and shovel plates, and samples of ovals. This iron cannot fail to recommend itself to all interested in the manufacture or use of the metal. Its invention and manufacture yield a material possessing a surface nearly as dense as steel, much increased elasticity, and greater resistance to tensile and torsional strains than the same sectional area of iron finished in the ordinary method. The bars are condensed in the finishing process to their centres, as will appear from the subjoined experiments, made by Merrick & Sons, at their Southwark Foundry.

### TENSILE STRENGTH.

Lbs. per sq. in.

Sample No. 1, inferior quality, broke at................ ........49,510
Same bar polished and condensed.................................66,862

Sample No. 2 broke at........................................57,350
Same bar polished and condensed.............................92,623

35,273—increase, .61

### TORSIONAL STRENGTH.

Sample No. 3, $1\frac{5}{16}$ diameter; twisted at a strain of $587\frac{1}{2}$ lbs. on a lever 25 inches long.

Same bar polished and condensed, $1\frac{1}{4}$ diameter; twisted at a strain of 1,000 lbs. on a lever 25 inches long. Increase, $413\frac{1}{2}$ lbs.=.97.

*A First Class Premium.*

Finally, the makers report results of tests made on cold-rolled iron finger bars for mowing machines, placed in competition with steel, thus:

## COMPARATIVE TEST OF FINGER BARS.

### STATEMENT No. 1,

Of a test made at the American Iron Works, Pittsburgh, Pa., August, 1865, to determine the comparative stiffness and elasticity of Finger Bars made of Patent Cold Rolled Iron, and those made of Cast Steel, dimensions of the Bars being the same.

| KIND OF BARS. | Weight applied. | Deflection. | Permanent Set. | Difference. |
|---|---|---|---|---|
| | Lbs. | in. | in. | in. |
| Cold Rolled | 270 | 8 | 1-16 | |
| Cast Steel | 250 | 8 | 3-16 | $\frac{1}{8}$ |
| Cold Rolled | 375 | 12 | $\frac{1}{2}$ | |
| Cast Steel | 310 | 12 | $1\frac{5}{8}$ | $1\frac{1}{8}$ |
| Cold Rolled | 450 | 15 | $1\frac{1}{2}$ | |
| Cast Steel | 330 | 15 | $3\frac{1}{2}$ | 2 |
| Cold Rolled | 550 | 19 | $4\frac{5}{16}$ | |
| Cast Steel | 380 | 19 | 7 | $2\frac{11}{16}$ |

Weight taken by Dynamometer.

### STATEMENT No. 2,

Of a test made at the Reaper and Mower Works of Walter A. Wood, Hoosick Falls, N. Y., October 24, 1866, to determine the comparative stiffness and elasticity of Finger Bars made of Patent Cold-Rolled Iron, and those made of Steel, dimensions of the Bars being the same.

| KIND OF BARS. | Weight applied. | Deflection. | Permanent Set. | Difference. |
|---|---|---|---|---|
| | Lbs. | in. | in. | in. |
| COLD ROLLED | | 6 | | |
| Cast Steel, No. 1 | | 6 | | |
| Cast Steel, No. 2 | | 6 | | |
| COLD ROLLED | | 8 | | |
| Cast Steel, No. 1 | | 8 | 5-16 | 5-16 |
| Cast Steel, No. 2 | | 8 | 1-16 | 1-16 |
| COLD ROLLED | | 10 | 1-16 | |
| Cast Steel, No. 1 | NOT TAKEN. | 10 | 13-16 | $\frac{3}{4}$ |
| Cast Steel, No. 2 | | 10 | $\frac{3}{8}$ | 5-16 |
| COLD ROLLED | | 12 | $\frac{1}{8}$ | |
| Cast Steel, No. 1 | | 12 | $2\frac{1}{4}$ | $2\frac{1}{8}$ |
| Cast Steel, No. 2 | | 12 | 1 | $\frac{7}{8}$ |
| COLD ROLLED | | 14 | $\frac{1}{8}$ | |
| Cast Steel, No. 1 | | 14 | $3\frac{3}{4}$ | $3\frac{5}{8}$ |
| Cast Steel, No. 2 | | 14 | $2\frac{1}{8}$ | $2\frac{5}{8}$ |
| COLD ROLLED | | 16 | $\frac{3}{4}$ | |
| Cast Steel, No. 1 | | 16 | $5\frac{3}{4}$ | 5 |
| Cast Steel, No. 2 | | 16 | $4\frac{3}{8}$ | $3\frac{5}{8}$ |

Taken all together, these tests and investigations form a very valuable and almost encyclopedic collection of facts for the use of the engineer. R. H. THURSTON.

# REPORT

ON

## STRENGTH, ELASTICITY, DENSITY AND OTHER PROPERTIES

OF

# COLD-ROLLED SHAFTING

MADE BY THE

## AMERICAN IRON WORKS, PITTSBURGH, PA.

AND ON THE

## Untreated Iron from which it is made.

By PROF. R. H. THURSTON,

*Director of the Mechanical Laboratory Stevens Institute of Technology, Hoboken, N. J., 1877.*

---

## REPORT ON TESTS BY TENSILE STRESS.

The object of the investigation on which the following is a report in detail, was to determine the effect upon iron of the process of cold-rolling as practiced at the American Iron Works, Pittsburgh, and to compare the results of tests made on iron thus treated, with those obtained under exactly the same conditions from tests of other samples untreated, but off the same bar and originally of the same structure, chemical composition and physical properties, and thus to bring out all the characteristic properties in such a manner that valid conclusions may be drawn as to whether and to what degree, the process of cold-rolling is beneficial or detrimental to iron as a material of construction.

The specimens tested by tension were 53 in number, constituting three lots :

Lot No. 1—Consisted of thirty-five specimens, of five different sizes, and each designated by a letter as follows : A, $2\frac{9}{16}$ inches in diameter; B, $2\frac{1}{16}$; C, $1\frac{3}{8}$; D, $1\frac{1}{16}$; E, $\frac{43}{64}$; and A', $2\frac{7}{16}$; B', 2;

C', $1\frac{5}{16}$; D', 1 inch; and E', $\frac{5}{8}$. A to E are the marks and sizes of the untreated metal, and A' to E' the sizes to which they were reduced by the process of cold-rolling. There were three specimens of each size untreated, and three of each cold-rolled. There was also one specimen of each size of the annealed cold-rolled iron.

The length of these specimens was originally 40 inches. They were all tested as they came from the rolls, without subsequent reduction in the lathe.

LOT No. 2—Consisted of six specimens turned down in the lathe, from bars originally two inches in diameter, to the sizes A, $1\frac{3}{4}$ inches; B, $1\frac{1}{2}$; C, 1. There were two specimens of each size, one cold-rolled and the other untreated.

The length of the reduced part, between shoulders, was 24 inches.

LOT No. 3—Consisted of twelve specimens, all turned in the lathe from round bars two inches in diameter to the sizes A, $\frac{7}{8}$ inch; B, $\frac{3}{4}$; C, $\frac{5}{8}$; D, $\frac{1}{2}$; E, $\frac{3}{8}$; F, $\frac{1}{4}$ inch in diameter—two specimens of each size, the one untreated and the other cold-rolled.

All these specimens were 8 inches between shoulders.

Lots 1 and 2 were tested by me at the works of the Keystone Bridge Company, at Pittsburgh, Pa.

The testing machine was of the form known as the "Hydraulic Machine," and consisted simply of a hydraulic cylinder in which pressure was obtained by pumps driven from the shafting and gauged by a Show and Justice gauge. The machine was designed for very heavy work, and was not well calculated to bring out slight variations of resistance in the material while under strain. The friction of the machine was probably sufficient to cover such slight variations.

There being no device for measuring the elongations, the following expedient was resorted to : a steel scale was secured at one end to the specimen by means of a clamp-screw, so as to be about equi-distant from the chucks. A mark was made on the bar at each end of this scale; these initial lines were thus made exactly 20 inches apart on the unstrained specimen.

The stress was applied gradually and regularly, and as the test-piece stretched a scriber was frequently drawn across the bar at the free end of the scale, thus marking the extensions due to the successive increments of load. These extensions were afterwards carefully measured, and their amounts are given in the tables, where they are placed opposite the corresponding loads.*

This method of indicating and measuring the extensions is not mathematically correct, since the distances between the marks, as measured after rupture, do not represent with mathematical accuracy the true elongations.† The error, however, which varies from

*Where the extensions are not given in the tables, they were either too small to be measured by the instruments available, or the marks were not legible.

†The inaccuracy in the measurements referred to above is due to the fact that the increments of extension, indicated by the marks, do not correctly represent the elongations caused by the corresponding loads, but are too great by a variable quantity due to the stretch of the bar caused by the continually increasing loads.

Thus the error decreases with the load, being a maximum for the first increment and zero for the last.

The true extensions can be determined in the following manner:

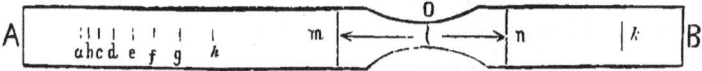

Let A B represent a bar ruptured at $o$; $a$ and $k$ are the initial marks on the same, and $b$ to $h$ marks indicating elongations.

Let $t_1 = a\,b$, the indicated extension for the first load.

$\quad t_2 = a\,c$,     "     "     "   second "

$\quad t_3 = a\,d$,     "     "     "   third "

$\quad t_4 = a\,n$,     "     "     "   any "

$T = a\,k - l$ (the original length) which is the true extension.

$T_1 =$ the true extension for the first load.

$T_2 =$   "     "     "   second load.

$T_n =$   "     "     "   any load.

$T_x =$ the difference between the length of the contracted portion $m\,n$ and a cylinder of the same volume, but equal in diameter to the uniformly reduced portion of the bar, and letting $l =$ the original distance between the initial marks.

Then we have :

$$T_1 = t_1 - T_1 \frac{T - T_x - T_1}{l} \text{ similarly}$$

$$T_2 = t_2 - T_2 \frac{T - T_x - T_2}{l} \text{ and}$$

$$T_n = t_n - T_n \frac{T - T_x - T_n}{l}.$$

By aid of these formulas the true extension may be found from the measurements. Although from scientific points of view such errors cannot be overlooked, yet practically they may be neglected. It has, therefore, been considered not necessary to make the corrections in the table.

a maximum in the first increment to zero in the last, is quite too insignificant to cause an appreciable change in the strain-diagram plotted from the tables, and may be neglected.

To avoid mistakes which might be caused by a misunderstanding of terms repeatedly used in this report, and to show how certain results given in the tables of comparison were derived, the following definitions and explanations will be found useful :

(1) The " Modulus of Elasticity " is the ratio of the elongation to the force which produces it, in a piece of which the length and cross-section are each unity.   Or, in other words, it is a force such as would cause a bar whose length and cross-section are unity to double its original length, provided the bar could be stretched so far and without change in the ratio of elongation.

If E denote the Modulus of Elasticity,

P, the force applied, (which force must not strain the specimen beyond its Elastic Limit,)

K=the area of cross-section,

L=the length of the bar,

$l$=the length of the extension for the load P.

Then

$$E = \frac{P\,L}{K\,l} \quad\dots\dots\dots\dots\dots\dots\dots\dots\dots\dots\dots\dots\dots (1)^*$$

Now, if K and L are each unity, then

$$E = \frac{P}{L},$$

as indicated by the first definition ; and if $l$=L, and K is unity, then

$$E = P,$$

as stated in the second definition.

In our calculations L and $l$ are in linear, and K in square inches.  P has always been taken as great as possible, care being taken, however, to keep well within the Elastic Limit, i. e., that point at which the extensions cease to bear a constant proportion to the load.

*Wood's Resistance of Materials, new edition, p. 5.

(2) "Resilience" is a measure of the capacity of a material to resist shock, and its value is equal to the amount of energy expended, or the "work" performed, in producing distortion or rupture. The "Elastic Resilience" is the energy expended, or work performed, in straining a material to its elastic limit, and the "Ultimate Resilience" is the energy expended, or work performed, in breaking it; it is always equal to the product of the average resistance of the material into the distance through which that resistance is overcome.

The Moduli of Resilience, Elastic and Ultimate, are, therefore, the amounts of work done upon a specimen of material whose length and cross-section are both unity, to produce the above effects respectively.

To obtain a measure of this quantity:

Let W denote the Modulus of Ultimate Resilience,

$P_m$=the mean force necessary to rupture the specimen, and K, L, $l$=the same as in formula (1),

Then we have

$$W = \frac{P_m l}{KL} \quad \dots\dots\dots\dots\dots\dots\dots\dots\dots\dots\dots (2)$$

If K and L are both unity, then

$W = P_m l$=the number of units of work, as defined in this case, inch-pounds, or by taking L and $l$ in feet, we have foot-pounds; in the following discussion and in the annexed tables the latter will be used, it being the more common unit of work.

In the appended curves, representing graphically the behavior of the several specimens under stress, one inch on the vertical scale represents 10,000 pounds of stress per square inch of cross-section, and one inch on the horizontal scale, 2.5 per cent. of extension.

The Moduli of Resilience were calculated from these strain-diagrams by means of a formula derived in the following manner:

Let A=the area of the curve in square inches,

    C=the total stretch in per cent.,

    O=the mean ordinate of the curve, in inches,

    S=the extreme abscissa,

    $P_m$=10,000 O=mean stress.

Then

$$\frac{C}{2.5}=S, \text{ and}$$

$$\frac{A}{S}=2.5\frac{A}{C}=O, \text{ but}$$

$$10,000 \, O=P_m \, \therefore \, P_m=25,000\frac{A}{C}$$

Now, since

$$W=\frac{CP_m}{100} \text{ by definition, we have}$$

$$P_m=100\frac{W}{C}$$

substituting this value of $P_m$ above, we have for the Ultimate Resilience or shock-resisting power :

$$W=250A$$

similarly for the Elastic Resilience, or power of resisting a blow without permanent change of form :

$$W'=250A'$$

A' being the area of the curve up to the Elastic Limit.

To determine the Modulus of Elasticity with scientific accuracy in specimens of such limited dimensions as those here reported upon, it is necessary to measure the elongations within the Elastic Limit to within one-thousandth (0.001) of an inch. But with the means of measurement available where these specimens were tested, such precise determinations could not be made. For the Moduli of Elasticity, therefore, as well as the Moduli of Elastic Resilience and for the true extensions, reference must be made to the results reported in the tests of the specimens comprising Lot No. 3, which were tested in the Mechanical Laboratory of the Stevens Institute of Technology.

# COMPARISON OF COLD-ROLLED

###### WITH

# UNTREATED IRON SHAFTING.

*Discussion of Results Obtained from Tests of Specimens Comprising Lots Nos. 1 and 2.*

## LOT NO. 1.

**No. 1133.** Specimen No. 1133 was 40 inches long and $2\frac{7}{16}$ inches in diameter, rolled cold from a bar $2\frac{9}{16}$ inches in diameter; it was placed in the machine and a gradually increasing stress applied, its amount being noted at intervals when the extensions were marked as given in the appended record sheet. The curve No. 1133, on Plate No. 1, shows the behavior of this specimen while under stress. It passed its Elastic Limit under a load of 59,450 pounds per square inch of cross-section, and broke with a stress of 302,400 pounds, giving it a Modulus of Rupture of 64,800 pounds. Its total extension was only 1.15 per cent. The fracture was somewhat granular; the area at the fracture was reduced to 96.14 per cent. of the original; its Modulus of Ultimate Resilience is 600 foot-pounds and its Modulus of Rupture per unit of area of fractured section is 67,400 pounds.

**No. 1134.** Specimen No. 1134 was of the same dimensions and was rolled cold from the same bar as the preceding; it was tested in the same manner. (See record sheet and curve No. 1134, Plate I.) This specimen passed its Elastic Limit under a stress of 59,-450 pounds per square inch of cross-section, and after extending 2.75 per cent. it broke under a load of 310,200 pounds, giving it a Modulus of Rupture by tension of 66,500 pounds. The fracture was quite granular. The area at the point of fracture was reduced to 91.38 per cent. of the original ; the Modulus of Ultimate Resilience is 1,672 foot-pounds, and the Modulus of Rupture per square inch of fractured section is 72,700 pounds.

**No. 1135.** Specimen No. 1135 was of the same dimensions and was treated in the same manner as the two preceding specimens. The stress was noted and the extensions marked, as shown in the record of tests of this bar; curve No. 1135 graphically represents its behavior under stress. This specimen did not pass its Elastic Limit until the stress had reached 62,000 pounds per square inch of cross-section, after which it extended quite rapidly until it broke at a load of 312,700 pounds, giving it a Modulus of Rupture of 67,000 pounds. The total extension was 4.35 per cent., and the fractured area was reduced to 89.05 per cent. of the original. The Modulus of Ultimate Rsilience is 2,770 foot-pounds. The fracture was very similar in appearance to the two preceding specimens, and the Modulus per square inch of fractured section is 75,300 pounds.

**No. 1136.** Specimen No. 1136 was of untreated common iron ; it was 40 inches in length and 2$\frac{6}{16}$ inches in diameter, and was tested in the same manner as the preceding specimen, (see record sheet and curve No. 1136, Plate I.) This specimen passed its Elastic Limit under a stress of 29,800 pounds per square inch of cross-section, after which it extended with some irregularity, and finally broke under a load of 239,500 pounds, giving a Modulus of Rupture of 46,900 pounds. Rupture took place outside of the initial marks, hence the total extension could not be accurately ascertained ; the part of the bar between the initial marks had extended 20.55 per cent. The section at the fracture was reduced to 68.45 per cent. of the original section ; the fracture was quite rough and lamellar, and showed a slightly granular structure. The Modulus of Resilience, due to the above extension, is 8,777 foot-pounds. The Modulus of Rupture per unit of area of fractured section is 67,800 pounds.

**No. 1137.** Specimen No. 1137 was a bar of untreated iron of the same dimensions as the preceding; it was tested in the same manner, and the results tabulated as before, (see record sheet for No. 1137 and curve of the same number, Plate I.) This specimen passed its Elastic Limit under a stress of 26,200 pounds per square inch of cross-section, after which it extended regularly and quite rapidly, finally breaking under a load of 239,500 pounds, thus giving a Modulus of Rupture of 46,900 pounds. The total extension was 26.25 per cent., and the Modulus of Ultimate Resilience is 11,081 foot-pounds. The section of fracture was reduced to 62.15 per cent. of the original, and the Modulus of Rupture per unit of fractured area is 74,700 pounds. The fracture was quite rough, almost ragged, and showed a granular structure quite plainly.

No. 1138. Specimen No. 1138 was of untreated iron of the same dimensions as the two preceding specimens ; it was tested in the same manner. The Elastic Limit was passed under a stress of 29,800 pounds per square inch of cross-section, and it finally broke under a load of 329,400 pounds, which gives a Modulus of Rupture in tension of 46,400 pounds. It broke outside of the initial marks, between which the extension was 18.20 per cent. The Modulus of Resilience for that extension is 7,710 foot-pounds. The fractured area was 67.91 per cent. of the original section, and the Modulus of Rupture per square inch of fractured area is 68,-500. The fracture is rough and lamellar, and plainly shows a granular structure.

No. 1139. Specimen No. 1139 was 40 inches long and $2\frac{7}{8}$ inches in diameter ; it was cold-rolled from a bar $2\frac{7}{8}$ inches, and was afterwards annealed and tested in the usual manner. (See records of tests of 1139 and curve No. 1139, Plate I.) Its Elastic Limit was passed under a stress of 31,400 pounds per square inch of cross-section, and broke under a load of 216,116 pounds, thus having a Modulus of Rupture of 46,300 pounds. It broke outside of the scale, having extended 14.25 per cent. between the initial marks. The Modulus of Resilience for that extension is 6,076 foot-pounds. The fracture was quite rough, and showed some signs of granular structure. The fractured area was 61.20 per cent. of the original. The Modulus of Rupture per square inch of fractured area is 75,400 pounds.

No. 1140. Specimen No. 1140 was 40 inches in length and 2 inches in diameter. It was cold-rolled from a bar $2\frac{7}{8}$ inches in diameter, and was tested in the same manner as the specimens already described. The extensions for the successive loads were marked as shown in the appended record of test, and a graphical representation of the behavior of the specimen under stress is given by the curve No. 1140, Plate II.

This specimen passed its Elastic Limit under a stress of 57,500 pounds per square inch of cross-section, after which it extended regularly and rapidly, and broke under a load of 208,500 pounds, giving it a Modulus of Rupture of 66,400 pounds. It broke outside of the scale, the extension between the initial marks being 5.60 per cent.; the section at the fracture was reduced to 72.26 per cent. of the original area. The fracture, although generally rough and fibrous, showed a slightly granular structure in some places. The Modulus of Resilience for the measurable extension is 3,107 foot-pounds, and the Modulus of Rupture per square inch of the fractured section is 91,800 pounds.

**No. 1141.** Specimen No. 1141 was of the same dimensions and cold-rolled from the same bar as No. 1140 ; it was tested in the same way. This bar was placed in the machine twice, giving way under a load of 211,100 pounds the first time, and sustaining 213,650 pounds before it broke at the second trial. Rupture occurred outside the scale each time, and the marks were not legible ; no curve could therefore be plotted. The Modulus of Rupture is 67,200 pounds, and the Modulus of Rupture per square inch ot fractured section is 83,500 pounds. The section at the fracture was reduced to 80.22 per cent. of the original section ; the fracture was laminated and somewhat granular. The Elastic Limit, the Modulus of Resilience and the extension could not be determined.

**No. 1142.** Specimen No. 1142 was of the same dimensions as No. 1141, and was treated and tested 'in precisely the same way. It passed its Elastic Limit under a load of 57,500 pounds per square inch of cross-section, after which it extended rapidly and regularly over 11.00 per cent., and finally broke outside of the scale under a stress of 211,100 pounds, thus having a Modulus of Rupture of 67,200 pounds. The Modulus of Resilience, determined from the extension between the initial marks, is 7,136 foot-pounds. The area of cross-section of fracture was 73.22 per cent. of the original ; the Modulus of Rupture per square inch of fractured section is 91,900 pounds. The fracture was rough and laminated and showed a slightly granular structure.

No. 1143. Specimen No. 1143 was of untreated iron, $2\frac{7}{10}$ inches in diameter, and of the same length as all the preceding; it was tested in the same way. (Consult the appended record of test and curve No. 1143, Plate II.) This specimen passed its Elastic Limit under a stress of 28,200 pounds per square inch of cross-section, then extended regularly and rapidly until it broke under a load of 162,200 pounds, having elongated 19.25 per cent. The Modulus of Rupture is, therefore, 48,500 pounds, and the Modulus of Ultimate Resilience is 8,232 foot-pounds. The area of the cross-section at fracture was reduced to 70.37 per cent. of the original. The fracture was quite rough, almost ragged, somewhat laminated, and finely fibrous in structure thoughout. The Modulus of Rupture per square inch of fractured section is 69,000 pounds.

No. 1144. Specimen No. 1144 was untreated iron, of similar dimensions with the preceding; the extensions were marked in the same manner. (See appended record sheets and curve No. 1144, Plate No. I.) Its Elastic Limit was passed under a load of 28,200 pounds per square inch of cross-section, but extended 28.75 per cent. before it broke, under a load of 163,400 pounds, which gives it a Modulus of Rupture of 48,900 pounds. The Modulus of Ultimate Resilience is 12,717 foot-pounds, and that of Rupture per unit of fractured section is 81,300 pounds. The area of the fractured section was reduced to 60.20 per cent. of the original; this fracture was rough but not ragged, and generally finely fibrous.

No. 1145. Specimen No. 1145 was in all respects like the two preceding. This specimen was placed in the machine twice, breaking unfairly each time; it broke under a load of 160,880 pounds on the first test, and sustained 177,650 pounds at the second trial. It was reduced from its original diameter, $2\frac{7}{10}$ inches, to $1\frac{13}{16}$ inches in diameter by the first test, and showed an increased strength with the reduced diameter. It broke in the jaws of the clamp at the second trial. The extensions could not be read. The Modulus of Rupture in the first test was 48,100 pounds. The fracture was quite rough and somewhat granular.

No. 1146. Specimen No. 1146 was 2 inches in diameter and of the standard length. It had been cold-rolled from a bar $2\frac{7}{10}$ inches in diameter to the above size, and was subsequently annealed. It passed its Elastic Limit under a stress of 31,800 pounds per square inch of cross-section, after which it extended regularly and more and more rapidly, as is shown by the curve No. 1146, Plate II, and finally broke under a load of 154,400 pounds, thus giving a Modulus of Rupture of 49,600 pounds. Rupture took place outside the scale. The extension between the initial marks was 12.50 per cent.; the Modulus of Resilience due to that elongation is 5,619 foot-pounds. The Modulus of Rupture per unit of area of fractured section is 78,400, and the area at the point of fracture was reduced to 63.22 per cent. of the original dimension. The fracture was rough, and very completely fibrous.

No. 1147. Specimen No. 1147 was of the standard length and 1⅛ inches in diameter. It was cold-rolled from a bar of untreated iron 1¾ inches in diameter. (See appended record of test and curve No. 1147, Plate III.) This specimen passed its Elastic Limit under a load of 56,200 pounds per square inch, which is a comparatively low figure; it then elongated very gradually until the stress had reached 64,500 pounds per square inch, after which it yielded more rapidly, and broke under a load of 91,400 pounds, having extended 5.85 per cent. The Modulus of Rupture in tension is 67,500 pounds, and the Modulus of Ultimate Resilience is 3,616 foot-pounds. The fractured section was reduced to 76.79 per cent. of the original, and the Modulus of Rupture per unit area of fractured section is 87,900 pounds. The fracture was rough, and completely and finely fibrous.

No. 1148. Specimen No. 1148 was in every respect like the preceding, and was tested in the same manner. It passed its Elastic Limit under a stress of 60,000 pounds per square inch of cross-section. After the load had been increased to 64,500 pounds per square inch it elongated very rapidly, and finally broke, after elongating 7.35 per cent., under a load of 91,400 pounds, which gives it a Modulus of Rupture of 67,500 pounds, and a Modulus of Ultimate Resilience of 4,710 foot-pounds. The area at the fracture was reduced to 70.22 per cent. of the original cross-section ; the Modulus of Rupture per square inch of fractured area is 96,200 pounds. The fracture was quite rough, but not ragged, and was completely and finely fibrous.

No. 1149. Specimen No. 1149 was in all respects like the two preceding specimens; it was tested in the same manner. Its Elastic Limit was passed under a load of 56,200 pounds, the same as that of No. 1147, after which it elongated more rapidly than the latter, as is seen by studying the curves of No. 1147 and No. 1149, Plate III; it broke under a load of 92,700 pounds, thus giving a Modulus of Rupture of 68,500 pounds. Rupture took place outside the scale, the extension between the initial marks being 4.90 per cent., and the Modulus of Resilience due to this extension is 2,913 foot-pounds. The area of cross-section at the fracture was reduced to 82.19 per cent. of the original; the Modulus of Rupture per square inch of fractured area is 83,300 pounds. The fracture is rough and generally fibrous, showing traces of granular structure.

**No. 1150.** Specimen No. 1150 was of the same length as the preceding specimens; it was cut from a bar of common iron 1⅜ inches in diameter, and was not cold-rolled. It was tested in the usual manner, and the elongations marked as before. It passed its Elastic Limit under a stress of 24,300 pounds per square inch of cross-section, after which it extended rapidly, and finally broke outside of the scale under a load of 74,600 pounds, which gives it a Modulus of Rupture of 50,300 pounds. The extension between the initial marks was 22.00 per cent., and the Modulus of Resilience due to that extension is 92.72 foot-pounds. The area of the section at fracture was reduced to 60.37 per cent. of the original; the Modulus of Rupture per square inch of fractured section is 83,000 pounds. The fracture was quite rough and slightly laminated, and completely fibrous.

**No. 1151.** Specimen No. 1151 was in every respect like the preceding, and tested in the same manner. It passed its Elastic Limit under a load of 24,300 pounds per square inch of cross-section, then extended regularly and rapidly until it broke under a load of 74,600 pounds, giving a Modulus of Rupture of 50,300 pounds. Rupture occurred outside the scale, the elongation between the initial marks being 15.65 per cent.; the Modulus of Resilience due to that extension is 6,497 foot-pounds. The area of section at the fracture was reduced to 62.81 per cent. of the original; the Modulus of Rupture per unit of area of fractured section is 80,000 pounds. The fracture was rough and laminated, showing a fine fibre with traces of a crystaline structure.

**No. 1152.** Specimen No. 1152 was like the preceding specimens, and it was tested in the same way. (See appended record of test and curve of No. 1152, Plate III.) This specimen passed its Elastic Limit under the same load as the other two, 24,300 pounds per square inch of cross-section, after which it elongated quite regularly until it finally gave way under a load of 74,600 pounds, giving as a Modulus of Rupture 50,300, which is the same as the two preceding specimens. Rupture took place outside the scale, the elongation between the initial marks being 22.60 per cent.; the Modulus of Resilience due to that extension is 10,141 foot-pounds. The area at the fractured section was reduced to 61.69 per cent. of the original; the Modulus of Rupture per square inch of fractured section is 81,500 pounds. The fracture was rough, and finely fibrous.

**No. 1153.** Specimen No. 1153 was of the standard length, and was 1ⁱ⁄₆₆ inches in diameter; cold-rolled from an untreated bar 1⅜ inches in diameter, but it was annealed before testing. It passed its Elastic Limit under a stress of 31,600 pounds per square inch of cross-section, after which it extended with some irregularity (see curve of No. 1153, Plate III), and finally broke under a load of 65,400 pounds, having extended 9.50 per cent. The Modulus of Rupture is therefore 49,500 pounds, and the Modulus of Ultimate Resilience is 4,927 foot-pounds. The area of cross-section at the fracture was reduced to 56.91 per cent. of the original cross-section, and the Modulus of Rupture per square inch of fractured section is 86,900 pounds. The fracture was dark, rough and fibrous.

**No. 1154.** Specimen No. 1154 was of the standard length, and was cold-rolled to a diameter of 1 inch from a bar of untreated iron 1ⁱ⁄₆₆ inches in diameter; it was tested in the usual manner. It passed its Elastic Limit under a stress of 58,700 pounds per square inch of cross-section (see curve of No. 1154, Plate IV,) and broke after extending 8.05 per cent., under a load of 53,300 pounds, thus giving it a Modulus of Rupture of 67,800 pounds, and its Modulus of Ultimate Resilience is 5,164 foot-pounds. The area of section at rupture was reduced to 67.23 per cent. of the original, and the Modulus of Rupture per square inch of fractured section is 100,900 pounds. The fracture was rough and fibrous.

**No. 1155.** Specimen No. 1155 was in every respect the same as the preceding, and was tested in the same manner. It passed its Elastic Limit under a stress of 63,700 pounds per square inch of cross-section, after which it extended very rapidly (see curve of No. 1155, Plate IV); after it had extended a little over 2 per cent. the maximum load, 53,800 pounds, was applied, under which it elongated until it broke. The Modulus of Rupture was 68,500 pounds. Rupture occurred outside the scale; the elongations between the initial marks was 7.45 per cent., and the Modulus of Resilience due to that extension is 4,909 foot-pounds. The area of section at the fracture was reduced to 67.23 per cent. of the original, and the Modulus of Rupture per square inch of fractured section is 101,900 pounds. The fracture showed nothing peculiar.

**No. 1156.** Specimen No. 1156 was of standard length, of the same diameter as Nos. 1154 and 1155, and was tested in the same manner. It passed its Elastic Limit under a stress of 58,700 pounds per square inch of cross-section, after which it elongated pretty rapidly under an increasing resistance until it broke under a load of 53,500 pounds, having a Modulus of Rupture of 68,200. Rupture took place outside the scale; the extension between the initial marks was 5.30 per cent., and the Modulus of Resilience due to that extension is 3,360 foot-pounds. The area of section at the fracture was reduced to 67.23 per cent. of the original, which is the same amount as was noted with the two preceding specimens. The Modulus of Rupture per unit of fractured section is 101,400 pounds. The fracture was like those of the two preceding specimens.

**No. 1157.** Specimen No. 1157 was of standard length and 1⅞ inches in diameter. It was common iron and was tested in the usual way. This specimen passed its Elastic Limit under a load of 26,100 pounds per square inch of cross-section and finally broke outside the scale under a load of 42,000 pounds, which gives it a Modulus of Rupture of 47,300 pounds. The extension between the initial marks is 21.75 per cent., and the Modulus of Resilience due to that extension is 7,946 foot-pounds. The area of section at the fracture was reduced to 58.09 per cent. of the original area; the Modulus of Rupture per square inch of fractured section is 81,500 pounds. The fracture was rough but not ragged, and perfectly fibrous in its structure; several partial fractures appeared in the reduced portion of the specimen as though it had had an equal tendency to break in either of several places.

**No. 1158.** Specimen No. 1158 was similar in all respects to the preceding, and was tested in the same manner. It passed its Elastic Limit under a stress of 28,900 pounds per square inch of cross-section, after which it extended with considerable regularity, but with a much higher resistance than its two companion specimens* (see curves, Plate IV); its ultimate resistance is, however, only equal to one of them and less than the other. The Modulus of Rupture is 47,300 pounds. Rupture took place outside the scale, the elongation between the initial marks being 23.60 per cent.; the Modulus of Resilience due to this extension is 9,882 foot-pounds. The area of the section at the fracture was reduced to 61.03 per cent. of the original area; the Modulus of Rupture per square inch of fractured section is 77,600 pounds. The fracture was quite rough, almost ragged, and exhibited a fibrous structure.

**No. 1159.** Specimen No. 1159 was like the two preceding specimens and was similarly tested. It passed its Elastic Limit under a stress of 28,100 pounds per square inch of cross-section and finally broke outside the scale under a load of 42,200 pounds, having a Modulus of Resistance of 47,600 pounds. The extension between the initial marks was 25.30 per cent., and the Modulus of Resilience due to that extension is 9,718 foot-pounds. The area of section at fracture was reduced to 59.55 per cent. of its original measure; the Modulus of Rupture per unit of fractured area is 79,900 pounds. The fracture was the same as that of the preceding specimen.

**No. 1160.** Specimen No. 1160 was 1 inch in diameter, cold-rolled from a bar $1\frac{1}{16}$ inches in diameter; it was annealed before testing. It passed its Elastic Limit under a load of 32,700 pounds per square inch of cross-section, after which it extended

---

* In the curves Ncs. 1157 and 1159, exhibiting the behavior of these specimens, we have only one observation between the Elastic Limit and point of rupture; they may therefore not correctly represent the resistances of these specimens, although the initial part of the curve shows a weaker metal in both cases.

somewhat irregularly (see curve of 1160, Plate IV), and finally broke under a load of 39,900 pounds after extending 12.65 per cent. The Modulus of Rupture is 50,900 pounds and the Modulus of Ultimate Resilience is 5,857 foot-pounds. The area of the section at the fracture was reduced to 67.23 per cent. of the original area; the Modulus of Rupture per square inch of fractured section is 75,600 pounds. The fracture exhibited no peculiarities, but was finely grained and fibrous.

**No. 1161.** Specimen No. 1161 was $\frac{5}{8}$ inch in diameter; it was cold-rolled from a bar of common iron $\frac{43}{64}$ inch in diameter. It passed its Elastic Limit under a stress of 63,800 pounds per square inch of cross-section, after which it elongated more and more rapidly (see curve of No. 1161, Plate V), until it broke under a load of 22,600 pounds; the total extension being 4.85 per cent. The Modulus is 73,800 pounds per square inch, which is very much higher than was given by any of the preceding sizes. The Modulus of Ultimate Resilience is 3,566 foot-pounds. The area of section at the fracture is reduced to 69.10 per cent. of the original section; the Modulus of Rupture per square inch of fractured area is 106,800 pounds. The fracture was slightly ragged and showed signs of a granular structure.

**No. 1162.** Specimen No. 1162 was of the same dimensions and materials as the preceding, and was tested in the same manner. It passed its Elastic Limit under a load of 67,100 pounds, after which it extended very rapidly, and broke under a stress of 22,136 pounds, giving a Modulus of 72,200 pounds. The total extension was 3.75 per cent., and the Modulus of Ultimate Resilience is 2,657 foot-pounds. The area of cross-section at the fracture was reduced to 80.19 per cent. of the original, and the Modulus of Rupture per square inch of fractured section is 92,000 pounds. The fracture had the same appearance as that of the preceding specimen.

**No. 1163.** Specimen No. 1163, which was in all respects like the two preceding specimens, passed its Elastic Limit under a stress of 60,600 pounds per square inch of cross-section, having elongated 5.00 per cent. The Modulus of Rupture is 75,500 pounds, and the Modulus of Ultimate Resilience is 3,670 foot-pounds. The area of section at the fracture was reduced to 72.04

per cent. of the original ; the Modulus of Rupture per unit of fractured area is 104,800 pounds. The fracture was rough and almost entirely fibrous.

**No. 1164.** Specimen No. 1164 was $\frac{43}{64}$ inch in diameter and of untreated iron. It passed its Elastic Limit under a stress of 29,-200 pounds per square inch of cross-section, after which it elongated very gradually at first until the load was increased to 39,000 pounds per square inch ; it then yielded quite rapidly, (see curve of No. 1164, Plate V,) and finally broke under a load of 17,800 pounds, thus giving a Modulus of Rupture of 50,100 pounds. Rupture occurred outside the scale ; the extension between the initial marks was 16.55 per cent., and the Modulus of Resilience due to that extension is 8,001 foot-pounds. The area of the fractured section was reduced to 69.39 per cent. of the original ; the Modulus of Rupture per unit of fractured section is 72,200 pounds. The fracture was rough and fibrous.

**No. 1165.** Specimen No. 1165 was similar to the preceding, and was tested in the same manner. It passed its Elastic Limit under a stress of 29,200 pounds per square inch of cross-section, after which it extended under an increasing load with some regularity until its maximum resistance of 43,600 pounds per square inch was reached, with an extension of 7.35 per cent., (see curve of No. 1165, Plate V,) under which load it finally gave way, having elongated 19.35 per cent. The Modulus of Rupture of this specimen is only 43.600 pounds—much lower than either of its companion specimens ; the Modulus of Ultimate Resilience is 8,761 foot-pounds. The area of cross-section at the fracture was reduced to 62.34 per cent. of its original section ; the Modulus of Rupture per unit of fractured section is 60,900 pounds. The fracture was fibrous and slightly ragged.

**No. 1166.** Specimen No. 1166 was like the two preceding specimens, and was tested in the same way. It passed the Elastic Limit under a stress of 29,200 pounds per square inch of cross-section—the same as the two preceding specimens—after which it extended quite rapidly and regularly, but in a less ratio, as the maximum load, 17,900 pounds, was approached, (see curve of No. 1166, Plate V,) which was reached with an extension of 17.50 per cent.; the resistance then decreased until the specimen broke under a load of 17,300 pounds. The Modulus of Rupture of this specimen is 50,800, which is higher than that of any other specimen of untreated iron in this series. Rupture took place outside the scale; the extension between the initial marks was 20.95 per cent. The Modulus of Resilience due to that elongation is 10,032 footpounds. The area of the cross-section at the fracture was reduced to 62.34 per cent. of its original amount; the Modulus of Rupture per unit of fractured section is 78,500 pounds. The fracture was similar to that of the preceding specimen.

**No. 1167.** Specimen No. 1167 was of the standard length and $\frac{3}{4}$ inch in diameter. It was cold-rolled from a bar $\frac{13}{16}$ inch in diameter, and annealed before testing. This specimen passed its Elastic Limit under a stress of 33,600 pounds per square inch of crosssection, after which it extended rapidly and quite regularly, reaching its maximum resistance, 15,400 pounds, with an extension of 13.35 per cent.; it finally broke under a load of 14,900 pounds. The Modulus of Rupture was 48,700 pounds. Rupture occurred outside the scale. The extension between the initial marks was 15.80 per cent.; the Modulus of Resilience due to that extension is 6,777 foot-pounds. The area of section at the fracture was reduced to 64.00 per cent. of its original area, and the Modulus of Rupture per unit of fractured section is 76,100 pounds. The fracture was rough and fibrous; the surface showed a seam which was open in several places.

# COMPARISON

# TURNED IRON, COLD-ROLLED AND UNTREATED.

*Samples Turned from 2-Inch Bars.*

## DETERMINATION

OF THE

## EFFECT OF COLD-ROLLING UPON THE INNER PORTION OF THE BAR.

## LOT No. 2.

All the specimens in this lot were tested in the same manner as those of the preceding.

**No. 1168.** Specimen No. 1168 was prepared from a cold-rolled bar, 2 inches in diameter, by reducing its diameter in a turning lathe to 1¾ inches. The distance between shoulders was, in this set, 24 inches. This specimen passed its Elastic Limit under a stress of 63,900 pounds per square inch, after which it yielded very rapidly and broke under a load of 160,900 pounds, which gives it a Modulus of Rupture of 66,900 pounds. Rupture took place outside of the scale, the extension between the initial marks being 6 per cent.; the Modulus of Resilience due to that Elongation is 3,877 foot-pounds. The area of section at the fracture was reduced to 70.56 per cent of the original section ; the Modulus of Rupture per unit of fractured area is 94,800 pounds. The fracture was quite rough and fibrous, with some indications of a granular structure.

**No. 1169.** Specimen No. 1169 was prepared from the same 2-inch cold-rolled bar, and was turned down to a diameter of 1½ inches. This specimen passed its Elastic Limit under a stress of 56,600 pounds per square inch of cross-section, after which it extended irregularly (see curve of No. 1169, Plate VI), and finally broke under a load of 121,000 pounds, having extended 7.65 per cent. The Modulus of Rupture is 68,500 pounds, and the Modulus of Ultimate Resilience is 4,930 foot-pounds. The area at the fractured section was reduced to 71.70 per cent. of the original area ; the Modulus of Rupture per unit of fractured section is 95,500 pounds. The fracture was rough and fibrous, with a partly granular structure.

No. 1170. Specimen No. 1170 was prepared by turning the 2-inch bar down to a diameter of 1 inch. This specimen passed its Elastic Limit under a stress of 56,700 pounds per square inch of section, after which it yielded quite rapidly and broke under a total load of 47,600 pounds, having extended 6.55 per cent. The Modulus of Rupture is 60,600 pounds; the Modulus of Ultimate Resilience is 3,794 foot-pounds. The area at the fractured section was reduced to 68.88 per cent. of the original dimension ; the Modulus of Rupture per square inch of fractured area is 86,200 pounds. The fracture was rough and fibrous, plainly showing in some places a granular structure.

No. 1171. Specimen No. 1171 was prepared from a bar of untreated iron, $2\frac{1}{16}$ inches in diameter, and was turned down to a diameter of $1\frac{3}{4}$ inches ; the distance between shoulders was made 24 inches, as before. This specimen passed its Elastic Limit under a stress of 30,900 pounds per square inch of cross-section, after which it elongated gradually until it had nearly reached its maximum resistance ; it then rapidly yielded (see curve of No. 1171, Plate VI), and finally broke under a load of 117,100 pounds, having extended 30 per cent. The Modulus of Rupture is 48,700 pounds, and the Modulus of Ultimate Resilience is 14,120 foot-pounds. The area at the fractured section was reduced to 58.62 per cent. of the original section ; the Modulus of Rupture per square inch of fractured section is 83,100 pounds. The fracture was rough and fibrous.

No. 1172. Specimen No. 1172 was prepared from a bar of untreated iron, $2\frac{1}{16}$ inches in diameter, by being turned down to a diameter of $1\frac{1}{2}$ inches. This specimen passed its Elastic Limit under a stress of 33,500 pounds per square inch of cross-section, after which it extended quite rapidly and pretty regularly until it finally broke under a load of 87,420 pounds, having extended

25.70 per cent. The Modulus of Rupture is 49,500 pounds, and the Modulus of Ultimate Resilience is 11,567 foot-pounds. The area at the fractured section was reduced to 59.82 per cent. of the original; the Modulus of Rupture per unit of fractured area is 82,700 pounds. The fracture was rough and fibrous, with signs of granular structure.

**No. 1173.** Specimen No. 1173 was prepared from the same bar of untreated iron, 2⅞ inches in diameter, by being turned down to a diameter of one inch. This specimen passed its Elastic Limit under a stress of 26,000 pounds per square inch of cross-section, after which it elongated more and more rapidly, and finally broke under a total load of 37,580 pounds, which gives it a Modulus of Resistance of 47,900 pounds. Rupture occurred outside of the scale, the extension between the initial marks being 21.30 per cent., and the Modulus of Resilience due to that elongation being 8,997 foot-pounds. The area of the fractured section was reduced to 60.86 per cent. of the original; the Modulus of Rupture per unit of fractured area is 78,700 pounds. The fracture was rough and fibrous.

## LOT No. 3.

Specimens No. 1104 (A, B, C, D, E, F,) and No. 1105 (A, B, C, D, E, F,) were tested in the Mechanical Laboratory of the Stevens Institute of Technology. The machine used is a Riehli Bros. Testing Machine.

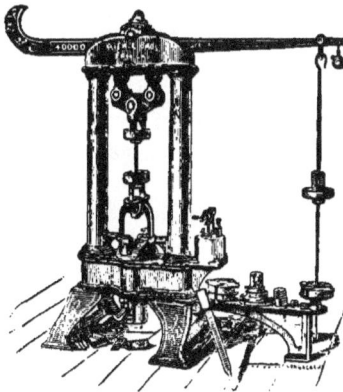

It consists of two strong cast iron columns secured to a massive bed-frame of the same material ; above these columns is fastened a heavy cross-piece, also of cast iron, containing two sockets, in which rest the knife edges of a large scale-beam. The upper chuck is suspended by two eye-rods from two knife-edges, $\frac{1}{2}$ inch to one side of centre of a heavy wrought iron block, which is hung by two links from two pairs of knife-edges, projecting from the scale-beam on opposite sides of the knife-edges of the latter and at equal distances from them, the whole making a very powerful differential beam-combination. All the knife-edges are of tempered steel, and the sockets and eyes are lined with the same material, thus reducing friction to a minimum. The load is applied by means of a hydraulic press with a fixed plunger and movable cylinder ; to the latter the lower chuck is fastened by means of an adjustable staple and link. The stress to which the test-piece is subjected is measured by means of suspended weights and a sliding poise (the latter not seen in the engraving.) The specimen is secured in the chucks either by wedge-jaws or cored chucks, according to the specimen to be tested. The capacity of the machine is twenty tons.

The extensions were measured by means of an instrument, in which contact is indicated by an "electric contact apparatus." The instrument consists essentially of two very accurately made micrometer-screws, working snugly in nuts, secured in a frame which is fastened to the head of the specimen by a screw-clamp ; it is so shaped that the micrometer-screws run parallel to, and equidistant from the neck of the specimen on opposite sides. A similar frame is clamped to the lower head of the specimen, and from it project two insulated metallic points, each opposite one of the micrometer-screws. Electric connection is made between the two insulated points and one pole of a voltaic cell, and also between the micrometer-screws and the other pole. As soon as one of the micrometer-screws is brought in contact with the opposite insulated point a current is established, which fact is immediately revealed by the stroke of an electric bell placed in the circuit. The pitch of the screws is 0.02 of an inch, and their heads are divided into 200 equal parts ; hence a rotary advance of one division on the screw-head produces a linear advance of one ten-thousandth (0.0001) of an inch. A vertical scale divided into fiftieths of an inch is fastened to the frame of the instrument and set very close to each screw-head and parallel to the axis of the screw; these serve to mark the starting point of the former, and also to indicate the number of revolutions made. By means of this double instrument, the extensions can be measured with great certainty and precision; and irregularities in the structure of the material causing one side of the specimen to stretch more rapidly than the other, do not diminish the accuracy of the measurements, since half the sum of the extensions indicated by the two screws is always the true extension caused by the respective loads.

Every possible precaution was taken during these tests to prevent the introduction of errors; and several special expedients were adopted to make the tests as complete as possible, with the object to bring out all characteristic properties.

Before beginning the tests, each specimen was very thinly covered with a coating of white lead, on which were marked the following lines:

(1.) Two straight lines on opposite sides of the specimen.

(2.) Eight lines at right angles to the former and exactly one inch apart, dividing the neck of each test-piece into seven equal cylinders, each of which was one inch long.

(3.) Lastly, a helix, of 1.56 inches pitch, was marked on each specimen the whole length of its turned part.

By means of these expedients irregularities of extension can be very readily observed and measured. An abrupt change of extension would distort the helix, thus indicating the precise spot where it occured. The originally equi-distant encircling lines served as measure of unequal extensions, and the lines parallel to the axis were useful as guides in taking these measurements correctly.

Fig. 1, Plate 16, represents specimens Nos. 1104$A$ and 1005$A$ before the test, and Fig. 2 shows No. 1105$A$, and Fig 3 No. 1104$A$ after the test.

In the sketches the specimens are represented half size.

No. 1105A. Specimen No. 1105$A$ was cold-rolled. Its length between fillets was 7.65 inches; its diameter was 0.875 inch. The specimen was secured in the chucks by wedge jaws, and the stress applied in increments of a thousand (1,000) pounds; the extensions were noted for every load, as is seen in the record sheet of test of No. 1105*A*. It passed its Elastic Limit under a stress of 54,900 pounds per square inch of cross-section, after which it elongated more and more rapidly, as can be best seen by studying the curve* of 1105*A*. It finally ruptured under a load of 39,600 pounds, or 65,850 pounds per square inch of cross-section, and 96,000 pounds per square inch of fractured area. The total elongation was 0.775 inch, or 11.07 per cent., which was distributed along the bar as follows :

Cylinder *a* extended 0.065 inch.
" *b* " .07 "
" *c* " .08 "
" *d* " .07 "
" *e* " .095 "
" *f* " .30 "
" *g* " .095 "

Total extension, - - 0.775 inch.

The portion *a b c d* had reduced to 0.84 inch, (*a* slightly more, *c* slightly less,) in diameter ; *e* and *g* tapered toward *f*, in the middle of which part it broke. The diameter of the fractured area was reduced to 0.725 inch. The fracture was rough, but not ragged ; it had a silky texture, and showed traces of granular structure. The surface of the specimen became slightly undulated, but did not exhibit the fibre, and showed no seams. This specimen had a Modulus of Elasticity of 26,871,000 ; its Modulus of Ultimate Resilience was 7,200 foot-pounds, and that of its Elastic Resilience 109.76 foot-pounds.

---

*NOTE.—The great extension for the load, 29,000 pounds, shown in the table, and more plainly seen on the curve, is due to the breaking of one of the wedges in the chuck. To provide against further accidents of this kind the specimens Nos. 1105 (A, B, C, D,) and Nos. 1104 (A, B, C, D,) had threads cut on both ends and a round nut screwed on, as shown in Fig. 1, Plate 16; this adapted the specimen to the shoulder jaws. Specimens E and F of each number were left plain, since the torsional stress necessarily brought on such slender pieces in cutting the threads might have strained them beyond the Elastic Limit, which would alter their resistance for tension, and thus render them of little or no value.

**No. 1104A.** Specimen No. 1104A was of untreated iron, and was of the same dimensions as the preceding. The stress was applied by increments of thousand pounds, and the extensions noted for each load were as shown in the appended record sheet. It passed its Elastic Limit under a stress of 14,000 pounds, or 23,300 pounds per square inch of cross-section, after which the elongations increased rapidly and quite uniformly, as is shown by the smoothness of the curve ; with 30,000 pounds of load, equivalent to 49-890 pounds per square inch of cross-section, the test was discontinued.\* The specimen had elongated 1.43 inches, or 20.43 per cent.; this elongation was exceedingly uniform, and the diameter quite uniformly reduced to 0.82 inch. Forty-eight hours afterward the specimen was again put in the machine and the test continued ; the stress was applied as before, up to 20,000 pounds, and then increased by increments of 2,000 pounds. Under a total load of 34,000 pounds, or 56,542 pounds per square inch of cross-section, it ruptured just below the nut in the upper chuck ; it was also cracked near the nut in the lower chuck. New threads were cut and the specimen finally ruptured at e, (Fig. 3,) under a total load of 35,150 pounds, equivalent to 58,450 pounds per square inch of original cross-section, and 122,452 pounds per square inch of fractured area. The total extension was 1.84 inches in 7 inches, or 26.30 per cent. This extension was distributed as follows :

Cylinder *a* extended 0.235 inch.
      "   *b*   "   .26   "
      "   *c*   "   .245  "
      "   *d*   "   .22   "
      "   *e*   "   .40   "
      "   *f*   "   .24   "
      "   *g*   "   .24   "
                         ————
Total extension,  -  -  1.84  inch.

The diameters of the portions *a b c d* and *f g* do not vary much from 0.782 inch. The diameter of the fractured area was 0.708 inch. The fracture differs from that of the preceding test piece

\*Note.—The specimen had extended beyond the range of the machine, so that a few changes had to be made before the testing could go on.

only in exhibiting a much more granular structure showing numerous clusters of irregular facets, especially near the edges of the fracture. The fracture at the end of the specimen exhibits the granular structure still more fully, nearly one-half of it being entirely of this character. This structure is often said to be indicative of a deficiency in ductility ; but the facts in the present case certainly show that this is, at least, not invariably the fact. It may here have been due to the intermitted test.

Judging from the appearance of the curve, and still more conclusively from the fact that it gave way in the chucks, at a part which had been but slightly affected by the stress brought upon it during the first trial, under a load of 53,444 pounds per square inch of the original cross-section of that part, it is evident that this specimen would very probably not have resisted much more than 50,000 pounds per square of such cross-section could it have been broken at the first pull. After resting two days its molecules seem to have had time to rearrange themselves, and thus to have made the specimen stronger than before. This is indicated in the sudden rise seen in the curve No. 1104A at a. (See Plate VII.)

Considering it as a new specimen having a diameter of 0.82 inch, it surpasses even the best cold-rolled specimen in this collection in ultimate strength, and it has considerably higher Elastic Limit ' but it falls below the average in ductility. Consulting the record of No. 1104A, or glancing at the curve No. 1104A', Plate VII, it is seen that it does not pass its Elastic Limit until after it has been strained above 60,000 pounds per square inch of cross-section, and that it bears over 66,800 pounds per square inch of cross-section before rupture takes place.

The Modulus of Elasticity of this specimen was only 23,860,000 ; its Modulus of Elastic Resilience was 14.91 foot-pounds, and its Modulus of Ultimate Resilience 12167.5 foot-pounds. , The latter is of course unduly augmented by the rest given the specimen after having been strained almost to its point of rupture.

I infer from the results of the peculiar treatment of this test piece, that the cold-rolling of this metal had not been carried far enough to produce the maximum effect attainable by the application of this process.

**No. 1105B.** Specimen No. 1105B was cold-rolled; it had a

diameter of 0.75 inch, and a length between fillets of 7.5 inches. The load was applied and the extensions noted as before. (See record sheet of 1105B.) Under a stress of 56,600 pounds per square inch of cross-section, it passed its Elastic Limit, the extensions increasing rapidly but regularly. It finally broke under a total load of 29,000, or 65,640 pounds per square inch of cross-section, and of 93,000 pounds per square inch of fractured area.

The total elongation was 0.63 inch in 7 inches, and was distributed as follows :

Cylinder $a$ extended 0.06 inch.
     "   $b$   "   .06  "
     "   $c$   "   .06  "
     "   $d$   "   .07  "
     "   $e$   "   .07  "
     "   $f$   "   .09  "
     "   $g$   "   .22  "

Total extension,   -   -   0.63 inch.

The diameters of the several parts $a\,b\,c\,d\,e$ were very nearly uniformly reduced to 0.72 inch. The diameter of the fractured section was rough and had a silky texture, and a small portion of it showed a granular structure. The surface was but very slightly undulated, and showed no fibre. The Modulus of Elasticity was 27,829,000. The Modulus of Elastic Resilience was 69.59 foot-pounds, and that of Ultimate Resilience was 5892.5 foot-pounds.

**No. 1104B.** Specimen No. 1104B was of untreated iron. Its dimensions were the same as those of the preceding. The stresses were applied and observed as before. (See appended record sheet of No. 1104B.)

This specimen passed its Elastic Limit under a stress of 23,800 pounds per square inch of cross-section, after which the elongations increased more and more rapidly but quite regularly (see curve). Under a load of 38,500 pounds per square inch it fairly began to "flow," stretching over 5 per cent. with an increment of 1,000 pounds of load ; it then again recovered its resisting power, as is indicated by the counterflexure of the curve. The next increment of 1,000 pounds of load produced an elongation of 1.5 per cent. only. The piece then extended more rapidly but regularly, and finally ruptured under a total stress of 21,800 pounds, or 49,330 pounds per square inch of original cross-section, and of 79,400 pounds per square inch of fractured area.

The total elongation was 1.51 inches in seven inches, or 21.57 per cent.; it was distributed as follows:

Cylinder *a* extended 0.12 inch.
"   *b*   "   .14 "
"   *c*   "   .13 "
"   *d*   "   .40 "
"   *e*   "   .305 "
"   *f*   "   .24 "
"   *g*   "   .175 "

Total extension,   -   1.51 inch.

The portion *a b c* had reduced to a diameter of 0.69 inch ; the fracture occurred in *d*, near *e*, as shown in fig. The diameter of the fracture was 0.592 inch ; it was rough, had a silky, fibrous texture, and showed no traces of granulation in structure. The surface of the cylindrical portion was undulated and exhibited a fibrous structure, especially near the fracture, where a few small seams also appeared.

The Modulus of Elasticity was 25,679,000. The Modulus of Elastic Resilience was 18.77 foot-pounds, and the Modulus of Ultimate Resilience was 8607.5 foot-pounds.

**No. 1105C.** Specimen No. 1105*C* was cold-rolled. Its diameter was 0.25 inch, and its length between fillets 7.45 inches. The stress was applied by increments of 800 pounds (see appended record of test). This specimen is remarkable for the wonderful regularity of its extensions with equal increments of stress ; the greatest variation in extension with the same increment of stress until 6,200 pounds total load is reached, was only 0.00,005 inch, (see record of No. 1105*C*.)

The Elastic Limit was fully passed under a total stress of 17,-000 pounds, or 55,400 pounds per square inch of cross-section.

After passing the Elastic Limit it elongated more and more rapidly, but very regularly, until it finally broke under a total stress of 20,450 pounds, or of 66,650 pounds per square inch of original cross-section, and 90,600 pounds per square inch of fractured area.

The total elongation was 0.645 inch in 7 inches, or 9.22 per cent., which was distributed as follows:

Cylinder *a* extended 0.06 inch.

| | | | | |
|---|---|---|---|---|
| " | *b* | " | .065 | " |
| " | *c* | " | .06 | " |
| " | *d* | " | .06 | " |
| " | *e* | " | .14 | " |
| " | *f* | " | .18 | " |
| " | *g* | " | .08 | " |

Total extension, - - 0.645 inch.

The portion *a b c d* was reduced very regularly, and measured 0.605 inch in diameter; the diameter of the fractured area was 0.536 inch. The fracture was very similar to the preceding; it had a fibrous, silky texture, and showed no trace of granular structure. The cylindrical surface became very slightly undulated under test, but revealed no seams; a trace of a seam appeared in *f*, near the fracture.

This specimen had a Modulus of Elasticity of 25,743,000; its Modulus of Elastic Resilience is 71.47 foot-pounds, and that of Ultimate Resilience is 56.40 foot-pounds.

**No. 1104C.** Specimen No. 1104*C* was of the same untreated iron; the dimensions were the same as those of the preceding specimen, and it was tested in the same manner, (see appended record of test.) It passed its Elastic Limit under a total stress of 7,400 pounds, or 24,000 pounds per square inch of cross-section. After passing the Elastic Limit, the "flow" of the molecules was twice retarded (see curve of No. 1104*C*, Plate VII,) the first time under a total load of 9,000 pounds, and the second time under 12,000. The extension then became regular and more and more rapid until rupture occurred under a total load of **15,500** pounds, or of 50,520 pounds per square inch of cross-section, and of 90-100 pounds per square inch of fractured area.

The total elongation was 1.72 inches in 7 inches, or 24.571 per cent., and was distributed as follows:

> Cylinder $a$ extended 0.185 inch.
> "    $b$    "    .2    "
> "    $c$    "    .2    "
> "    $d$    "    .22    "
> "    $e$    "    0.315    "
> "    $f$    "    **.**.34    "
> "    $g$    "    .26    "

> _____

> Total extension,    -    -    1.72 inches.

The portion $a\ b\ c\ d$, was reduced to 0.542 inch in diameter, $d$ decreasing toward $e$. Rupture took place between $e$ and $f$. The diameter of the fractured area was 0.48 inch; the fracture was rough, of a fibrous, silky texture, and showed decided indications of granular structure. The surface became quite undulated and exhibited a fibrous structure. The Modulus of Elasticity was somewhat high—30,363,000; the Modulus of Elastic Resilience was 18.09 foot-pounds and the Modulus of Ultimate Resilience was 10,875 foot-pounds.

**No. 1105D.** Specimen No. 1105$D$ was cold-rolled; it had a diameter of 0.5 inch and a length between fillets of 7.4 inches. The stress was applied in increments of 300 pounds and the extension noted for every addition of load, as before (see record of test). Although there is a disproportionate increment of extension with the increment of load passing from 5,000 to 5,300 pounds, as is noticeable, both in the curve and the record, it did not pass its Elastic Limit fully, until the total load, 11,000 pounds, or 56,800 pounds per square inch of cross-section, had been reached. The extensions then increased more and more rapidly; but no noticeable " flow " set in until the load had reached 61,000 pounds per square inch of cross-section. The specimen finally broke under a total stress of 13,000 pounds, or 66,200 pounds per square inch of cross-section, and 91,600 pounds per square inch of fractured area.

The total elongation in 7 inches was 0.57 inch, or 8.14 per cent., which was distributed as follows:

> Cylinder $a$ extended 0.055 inch.
> " $b$ " .06 "
> " $c$ " .06 "
> " $d$ " .065 "
> " $e$ " .07 "
> " $f$ " .19 "
> " $g$ " .07 "
>
> Total extension, - 0.57 inch.

The diameter of the fractured area was 0.425 inch, and that of the portion $a\ b\ c\ d$, was 0.484 inch ; $e$ and $g$ tapered slightly toward $f$, where rupture occurred. The fracture is rough, fibrous, and of a silky texture, and shows no traces of a granulation. The cylindrical surface was but very slightly undulated and shows no traces of fibre or seams. The Modulus of Elasticity was 28,989,000; the Modulus of Elastic Resilience was 71.47 foot-pounds and the Modulus of Ultimate Resilience was 5072.5 foot-pounds.

No. 1104D. Specimen No. 1104$D$ was of untreated iron, and had the same dimensions as the preceding. The stress was applied with the same increments and elongations which were noted as before. (See records.) It passed its Elastic Limit under a stress of 4,700 pounds, or 23,900 pounds per square inch of cross-section, after which the elongation increased in the usual manner with three successive increments of stress ; but the next increment produced a disproportionately large extension, and the next a disproportionately small extension, as is shown by the counterflexure of the curve of No. 1104$D$, Plate VIII. The specimen then elongated rapidly but regularly at a slightly diminishing rate toward the close of the test, as is indicated by the slight upward concavity in that part of the strain-diagram. The specimen finally broke under a total stress of 10,000 pounds, or of 50,980 pounds per square inch of cross-section, and 93,600 pounds per square inch of fractured area.

The total elongation was 1.3 inches in 7 inches, which was distributed as follows:

$$
\begin{array}{llll}
\text{Cylinder} & a & \text{extended} & 0.14 \text{ inch.} \\
\text{``} & b & \text{``} & .15 \text{ ``} \\
\text{``} & c & \text{``} & .145 \text{ ``} \\
\text{``} & d & \text{``} & .145 \text{ ``} \\
\text{``} & e & \text{``} & .145 \text{ ``} \\
\text{``} & f & \text{``} & .34 \text{ ``} \\
\text{``} & g & \text{``} & .235 \text{ ``} \\
\end{array}
$$

Total extension, - - 1.3 inch.

The diameter of the portion $a\ b\ c\ d\ e$, was reduced to 0.469 inch; and that of the fractured section measured 0.369 inch. The fracture was rough, of a silky, fibrous texture; the surface was moderately undulated, but did not exhibit the fibrous structure, except in the portion drawn down, where also a small seam appeared.

The Modulus of Elasticity was 30,150,000; the Modulus of Elastic Resilience, 26.33 foot-pounds, and that of Ultimate Resilience 7757.5 foot-pounds.

**No. 1105 E.** Specimen No. 1105*E*, of cold-rolled iron, was 0.375 inch in diameter and 7.375 inches in length between fillets. The stress was applied in increments of 300 pounds, and the elongations were noted for every load. (See appended record.) This specimen shows some irregularities of resistance within the Elastic Limit, as is shown both by the records and by the curve. Between the loads 28,928 and 33,500 pounds per square inch of cross-section, the extensions for equal increments of load are disproportionately large. The Elastic Limit was passed under a total load of 6,000 pounds, or of 54,300 pounds per square inch of cross-section, after which the piece stretched very rapidly, and finally broke under a total load of 7,000 pounds, or of 63,400 pounds per square inch of cross-section, and 99,000 pounds per square inch of fractured area.

The total elongation was 0.51 inch in 7 inches, or 7.286 per cent., which was distributed as follows:

Cylinder *a* extended 0.05 inch.
" *b* " .05 "
" *c* " .06 "
" *d* " .05 "
" *e* " .18 "
" *f* " .05 "
" *g* " .07 "

Total extension, - - .51 inches.

The diameter of the portion *a b c d* was reduced 0.36 inch, and the fractured section measured 0.3 inch. The fracture was rough, and of a fibrous and silky texture ; it showed traces of granulation. The cylindrical surface had become slightly undulating and showed a few slight seams in the portion *d e f* near the fracture.

The Modulus of Elasticity was but 22,261,000. The Modulus of Elastic Resilience was 163.13 foot-pounds, and that of Ultimate Resilience 4,247.5 foot-pounds. The irregularities of resistance of this specimen within the Elastic Limit very probably indicate defects in its internal structure, minute flaws. The low co-efficient of Elasticity, the low Modulus of Rupture, and also the low Modulus of Ultimate Resilience corroborate this supposition.

The Modulus of Elastic Resilience is exceptionally high, but it cannot be considered entirely reliable on account of the disproportionate elongations within the Elastic Limit.

**No. 1104E.** Specimen No. 1104*E* was of untreated iron of the same dimensions as the preceding, the stress being applied in the same manner and the elongations noted as before. (See appended record of tests.) This test-piece had fully passed its Elastic Limit with a total load of 2,800 pounds, or 20,800 pounds per square inch of cross-section. After passing the Elastic Limit it elongated with regularly increasing rapidity until the load became 33,500 pounds per square inch, when the extension was suddenly arrested. (See curve of No. 1104*E*, Plate VIII.) And with the next increment of 500 pounds—4,529 pounds per square inch of cross-section —the increment was not much greater than within the Elastic Limit for the same increment of load. An additional load of 50 pounds—440 pounds per square inch of cross-section—produced an elongation of 2.448 per cent., after which the "flow" was again suddenly arrested, the piece extending only 0.081 per cent., with an increment of 350 pounds—4,088 pounds per square inch of cross-cection. The next increment of 300 pounds produced an elongation of 6.189 per cent., after which "flow" was arrested a third time, and an increment of 200 pounds of stress was required to overcome this new molecular resistance. It then yielded rapidly, though considerably less in proportion to the load than any other specimen of this lot, as will be seen on examination of the inclination of the curve. It broke under a total stress of 5,800 pounds, or 52,540 pounds per square inch of cross-section, and 97,700 pounds per square inch of fractured area.

The sudden offsets in the curve show the peculiar behavior of this specimen very plainly.

The total elongation was 1.44 inches in 7 inches, or 20.57 per cent. ; it was distributed as follows :

Cylinder *a* extended 0.155 inch.
" *b* " .195 "
" *c* " .195 "
" *d* " .195 "
" *e* " .35 "
" *f* " .195 "
" *g* " .155 "

Total extension, - - - 1.44 inch.

The diameter of the portion *a b c d* was reduced to 0.343 inch and that of the fractured area measured only 0.275 inches. The fracture was rough, and generally of a silky texture; but it had decided traces of granular structure. The surface was undulated and indicative of a fibrous structure : a small seam appeared near the fracture.

The Modulus of Elasticity of this specimen was 27,038,000 ; the Modulus of Elastic Resilience was 26.25 foot-pounds, and that of Ultimate Resilience was 8,500 foot-pounds.

**No. 1105F.** Specimen No. 1105*F* was "cold-rolled;" its diameter was 0.25 inch, and its length between fillets 7.25 inches.

The stress was applied in increments of 400 pounds, and the extensions were noted for every load, as before (see record of No. 1105F). This specimen did not elongate regularly within the Elastic Limit; the variations, however, were too small to be noticeable in the curve; the Elastic Limit was passed under a load of 50,000 pounds per square inch of cross-section, after which it suddenly yielded (see curve of 1105F, Plate VIII) extending 1.96 per cent. with an increment of 200 pounds of load, or 5,074 pounds per square inch of cross-section. It then suddenly stiffened, elongating but 0.137 per cent. with the next increment of 200 pounds, and finally broke under a load of 8,175 pounds, equivalent to 64,660 pounds per square inch of cross-section, and 91,800 pounds per square inch of fractured area, having extended only 0.24 inch in 7 inches, or 3.43 per cent.

This elongation was distributed as follows :

Cylinder *a* extended 0.02 inch.
"       *b*    "    .02  "
"       *c*    "    .02  "
"       *d*    "    .02  "
"       *e*    "    .02  "
"       *f*    "    .025 "
"       *g*    "    .115 "

Total extension,   -    .24 inch.

The portion *a b c d e f* was reduced to 0.24 inch diameter, and the fractured area measured 0.21 inch in diameter. The surface remained almost entirely unaltered. The fracture was rough, had generally a silky, fibrous texture, and showed some traces of granular structure.

The Modulus of Elasticity was very high—35,553,000 pounds; the Modulus of Elastic Resilience was 51.163 foot-pounds, and that of Ultimate Resilience was 1,945 foot-pounds.

**No. 1104F.** Specimen No. 1104F was of untreated iron, and was of the same dimensions as the preceding. The loads were applied in increments of 400 pounds, and the elongations were noted for each load (see record of No. 1104F). This specimen, like the preceding, elongated irregularly within the Elastic Limit, which it passed at a total stress of 1,100 pounds—22,400 pounds per square inch of cross-section. After the Elastic Limit was passed, the piece elongated rapidly and quite regularly, until the load had been increased to 37,684 pounds per square inch of cross-section, at which point (see curve of No. 1104F) it began to yield rapidly, and broke under the comparatively low stress of 2,110 pounds, or 42,980 pounds per square inch of cross-section.

The total elongation was 1.185 inches in 7 inches, or 16.928 per cent., which was distributed as follows :

> Cylinder *a* extended 0.04 inch.
>       "   *b*    "    .155  "
>       "   *c*    "    .155  "
>       "   *d*    "    .155  "
>       "   *e*    "    .18   "
>       "   *f*    "    .235  "
>       "   *g*    "    .265  "
>
> Total extension,   -   1.185 inches.

The diameter of the portion *b c d* was 9.228 inch, and that of the fractured area, 0.185 inch. The fracture was rough, of a fibrous and silky texture, and showed but very slight indications of granulation. The cylindrical surface also showed the fibrous structure.

The Modulus of Elasticity was almost as high as in the preceding specimen, 33,317,000 pounds; the Modulus of Elastic Resilience was 23.52 foot-pounds, and that of Ultimate Resilience was 6,535 foot-pounds.

# CONCLUSIONS.

## I. THE UNTREATED IRON.

In reviewing the results which were discussed in the preceding pages, we find :

1st. That the untreated iron which was tested full size has an average breaking strength of 48,700 pounds per square inch of cross-section, and that the ultimate strength per unit of area of section decreases gradually as the diameter increases. Neglecting the size D, ($1\frac{1}{16}$ inches in diameter) which is exceptionally low, we get a tolerably smooth curve (BB, Lot II, Plate IX), which starting with the smaller size E' ($\frac{41}{64}$ inch diameter), having a Modulus of Rupture equal to 50,500 pounds per square inch of cross-section, the curve sinks very gradually, at first, toward the larger sizes, but more rapidly as it approaches the largest size A, ($2\frac{9}{16}$ inches diameter), of which the strength is only 46,700 pounds per square inch of cross-section, or 3,000 pounds below the average.

2d. That the Elastic Limit* is passed under an average stress of 27,600 pounds per square inch of cross-section, or 56 per cent. of its ultimate strength. The curve of average Elastic Limits deviates much more from a straight line than that of Ultimate Strengths, and does not run parallel with it, thus indicating that the Elastic Limit and the Ultimate Strength are not directly related.

---

\* The Elastic Limits as obtained from the tests made at the works of the Keystone Bridge Co., are considerably higher than those obtained from the specimens tested in the Mechanical Laboratory of the Stevens Institute of Technology. This will be seen on examining the tables, giving the summary of results, or comparing the curves C'C' and E'E', with G'G' and D'D', and A'A' with F'F'. Plate IX.

This difference, since it occurs to almost the same extent in the turned specimen as in those tested full size, is very probably not so much (or perhaps not at all) due to the superiority of the surface-metal as to the method of testing and the manner in which the extensions were marked, the mark denoting the extension for the corresponding load being made before the latter had time to produce the full effect on the specimens.

3d. That the total extension* averages about 25 per cent., and that the fractured section is reduced on an average to 63.5 per cent. of its original area.

4th. That the results obtained with the specimens, which were turned from a bar 2 inches in diameter to sizes varying from $\frac{1}{4}$ to $1\frac{3}{4}$ inches, compare favorably with those tested full size as they came from the rolls, the turned specimen having an average breaking strength of 49,500 pounds per square inch of cross-section, which is 800 pounds greater than that of the unprepared specimens, and 1,000 pounds greater than the average strength of the bar from which these specimens were prepared.

5th. That excepting the size F, ($\frac{1}{4}$ inch diameter) which, having a strength of 42,900 pounds per square inch only, is low and very probably test piece was unsound—the strength increases as the diameter decreases in a greater ratio than in the cases of the unturned specimens. (See curves, Plate IX.) This indicates that the interior portion of the bar is stronger[†] than the exterior.

6th. That the average Elastic Limit of Lot III is 23,000 pounds per square inch, or 45.7 per cent. of the ultimate strength, while that of Lot II is 30,000 pounds[‡] per square inch, or 61 per cent. of the ultimate strength, the latter being also higher than the average Elastic Limit of Lot I.

7th. That the average total extension of the turned specimens is quite as high as that of those unturned, and that it decreases with the diameter. Although there appears to be a tendency in the same direction in the unturned bars, it cannot be positively

* The true average per cent. of extension could not be determined, since only a few of the specimens broke between the initial marks. But the figure given above is not far wrong, and is on the safe side; the partial extensions, as measured between the initial marks, average about 20 per cent. The Modulus of Resilience being a direct function of the extension, is also slightly too small.

† The indicated increase of strength toward the centre of the bar is contrary to the general opinion of engineers, according to which the outer shell is the stronger metal. The above evidence being based on tests of but one specimen of each size, we are not justified in putting it forward as positively proven before it has been corroborated by further experiment.

‡ The difference between the Elastic Limits of Lots II and III is undoubtedly partly due to the method of testing; whereas, that between those of Lots I and II, can only be attributed to the material itself.

asserted, because but a few breaks occurred which give evidence on this point.

8th. That the reduction of section at the point of rupture was greater in the turned than in the unturned specimens, the average reduced area of the former being but 59.1 per cent. of the original, which is 4.4 per cent. less than that of the unturned specimen.

Consulting Table G, we find that although the total extension decreases with the diameter, the reduction of section increases, and in a much greater ratio.

9th. That the average Modulus of Elastic Resilience* is 21 31 foot-pounds, and that of Ultimate Resilience is 9,074 foot-pounds, the former being only 0.23 per cent. of the latter. Therefore, although the work necessary to be exerted in breaking a specimen of untreated iron may be very considerable, it requires only a very small fraction of that work to strain it beyond its Elastic Limit, i. e., to produce a permanent elongation which may be sometimes as serious a result as actual fracture.

## II. COLD-ROLLED IRON.

Summing up the results of tests obtained with cold-rolled iron, which were individually noticed in the preceding pages, we observe :

1st. That the specimens, which were tested of full size and just as they were taken from the rolls, have an average breaking strength of 68,600 pounds per square inch, and that this ultimate strength increases as the size decreases. (See curve of breaking loads AA, Lot I, Plate IX.) Beginning with the largest size, A, (2$\frac{1}{16}$ inches diameter,) of which the breaking load is 66,200 pounds per square inch, the ultimate strength increases regularly to size D, (1 inch diameter,) which has a Modulus of Rupture of 68,200 pounds. This is below the average breaking strength, which figure is carried above its proper amount by the exceptionally high resistance of E' ($\frac{3}{8}$ inch diameter,) which has an average Modulus of Rupture of 73,800 pounds.

---

*For Lots I and II, on account of the method of testing, the Modulus of Elastic Resilience could not be satisfactorily determined, and the Modulus of Ultimate Resilience being in most cases only *partial*, is not comparable with that of Lot III.

2d. That the average Elastic Limit is 59,600 pounds per square inch, nearly 87 per cent. of the ultimate strength.

Although the curve of Elastic Limits, A' A', is more regular than that of the untreated iron, C' C', it confirms the deduction already made that the Elastic Limit is not necessarily proportional to the ultimate strength of the material. The curves of Elastic Limits, E' E', G' G', F' F', D' D', and especially B' B', similarly indicate this independence; the latter curve is that of annealed cold-rolled iron, of which the breaking weight does not exceed that of the untreated iron ; it shows a much higher Elastic Limit than the latter.

3d. That the total extension is small in comparison with that of untreated iron ; it averages but 6 per cent. The extension of the largest size, A, (2$\frac{7}{15}$ inches diameter,) which was quite granular in structure, was but 2.75 per cent., and that of the smallest size, E, ($\frac{3}{8}$ inch diameter,) was 4.5 per cent. Size C (1$\frac{7}{15}$ inches diameter,) was the most ductile, elongating 8.3 per cent. In these cases the ductility does not decrease with diameter, as is the case with untreated iron.

4th. We find in comparing the results here obtained with those given by the specimens of Lots II and III, which were turned from the same cold-rolled bar, 2 inches in diameter, that the Modulus of Rupture is 65,300 pounds, which is 1,600 pounds less than that of the bar from which these specimens were prepared, and further, by studying the curve of breaking loads D D, E E,* it is seen that the Ultimate Strength gradually diminishes with the diameter, which shows that although the metal was not equally affected throughout the whole bar, the effect was very marked, even on the smallest size, of which the breaking load per square inch was only 2,200 pounds less than that of the full-sized bar, and was 16,200 pounds *greater than the Ultimate Strength of the untreated bar of the same iron.*

5th. That the average Limit of Elasticity of Lots II and III is

*The irregularities of the dotted curves of Lots II and III must not be attributed to the greater irregularities in the material which was turned to size than in that which was tested full size, but to the fact that each observation was obtained from a single test, and not from several, as in Lot I. If three or more specimens of the same size and material could have been tested, the curve would undoubtedly have been much smoother.

reached under a load of 57,000 pounds per square inch—that of Lot II at 59,000 pounds, and that of Lot III at 54,800 pounds per square inch. This difference, as already stated, is very probably partly due to the method of testing. The Elastic Limit of Lot III is 83.5 per cent. of its Ultimate Strength ; therefore, a structure of cold-rolled iron may be strained above 0.83 of its Ultimate Strength before it becomes injured by permanent distortion, whereas untreated iron reaches this stage with but 0.45 of its breaking load.

6th. That the average ductility of the turned specimens (Lots II and III) is about the same as that of the unturned (Lot I), or 6 per cent., but is less than that of the full-sized bar from which these specimens were prepared. The average elongations of Lot III separately, approaches that of the full-sized bar, but falls below,* the measurable extension of the latter being 8.3 per cent. The extension of Lot I appears to be independent of the diameter of the specimen, while in Lot III the elongation decreases with the diameter.

7th. That the reduction of area in the cold-rolled iron is much more nearly proportional to the extension than in the untreated iron, indicating that the process of cold-rolling renders the bar much more homogeneous.

8th. That the average Modulus of Resilience is 90.26 foot-pounds, which is 1.8 per cent. of the average Modulus of Ultimate Resilience, the latter being 5,000 foot-pounds. With untreated iron it was 0.23 per cent. The Modulus of Ultimate Resilience of the cold-rolled iron is 56 per cent. of that of the common iron ; but comparing the Modulus of Average Elastic Resiliences with each other, we see that they are to each other as 1:4.2 nearly, *i. e.*, it will require more than four times as much work to be expended to distort a structure of cold-rolled iron as will be required to strain one of similar proportions, but built of common iron.

9th. Comparing the fractures, we infer that cold-rolling does not produce crystalization ; many of the specimens, both cold-rolled and untreated, showed a granular structure plainly, and without exhibiting deficiency in ductility as a consequence.

*The total extension of the full-sized bar could not be ascertained, since all specimens broke outside of the marks

### III.  ANNEALED COLD-ROLLED IRON.

The results obtained with annealed cold-rolled iron show :

1st.  That the Average Ultimate Strength is reduced by anneal-ing in these cases to its original figure very nearly, and that its value increases as the diameter decreases, from 46,300 pounds per square inch, the Modulus of Rupture of size A′ (2$\frac{7}{16}$ inches diam-eter) to 50,900, the breaking load of size D′ (1 inch diameter). Size E′ ($\frac{5}{8}$ inch diameter) had a breaking load of only 48,700 pounds per square inch ; it very probably was flawy.

2d.  That the Average Elastic Limit is 32,000 pounds per square inch, considerably higher than that of the untreated iron, which rendered this limit at 27,600 pounds.

3d.  That the average total extension, 15 per cent.,* is interme-diate between that of the cold-rolled and that of the untreated iron ; it seems to be independent of the diameter of the specimens.

4th.  That the fractured section was reduced to 62.5 per cent. of its original diameter.  This reduction seems to be not only inde-pendent of the original diameter, but also independent of the total extension.  Size C′ extended only 9.5 per cent., while it shows a contraction of the fractured section to 57 per cent., whereas size D′ extended over 16 per cent., and had a fractured section of which the area was 67.25 per cent. of the original.

5th.  The Modulus of Ultimate Resilience, depending as it does upon the Elastic Limit, Ultimate Extension, and Modulus of Re-sistance, must be intermediate between that of cold-rolled and that of untreated iron, being greater than the former, and less than the latter.  Thus by annealing, the cold-rolled metal loses greatly in all its peculiarly valuable qualities, but retaining, nevertheless, a higher tenacity and higher Elastic Limit, with reduced ductility as compared with ordinary iron.

---

*The average total extension could not be exactly obtained, as many of the specimens broke outside of the marks.

# REPORT

O N

# TESTS BY TRANSVERSE STRESS

O F

## UNTREATED AND COLD-ROLLED

# WROUGHT IRON SHAFTING.

The object of the following investigation is the determination of the effect upon its transverse strength of cold-rolling wrought iron shafting of different diameters, comparing the cold-rolled metal with the same material hot-rolled.

The samples tested were made companion pieces to those constituting Lot I, tested by tensile stress, and reported upon in the preceding pages.

There were thirty-five samples in all, viz:

First, 15 of common or hot-rolled iron, comprising five sizes, nominally:* $A=2\frac{9}{16}$ inches in diameter ; $B=2\frac{1}{16}$ ; $C=1\frac{9}{16}$ ; $D=1\frac{1}{16}$ ; $E=\frac{43}{64}$. Secondly, 15 samples of cold-rolled iron, also of five sizes ; $A'=2\frac{7}{16}$ inches in diameter; $B'=2$ ; $C'=1\frac{5}{16}$ ; $D'=1$ ; $E'=\frac{5}{8}$. Thirdly, 5 samples of cold-rolled and annealed iron, one of each of the cold-rolled sizes.

The lengths of the test-pieces were 40 inches or more. In this report the sizes will always be designated by the letters A to E, as above.

---

*The diameters of the untreated iron, as obtained by actual measurement, and as given upon the record sheets, and upon which the calculations are based, do not exactly agree with the above. The nominal diameters of the cold-rolled bars, however, accord with the measurements made upon them.

## DESCRIPTION OF THE TRANSVERSE TESTING MACHINE.

The machine employed in testing by transverse stress is readily understood from the accompanying cut.

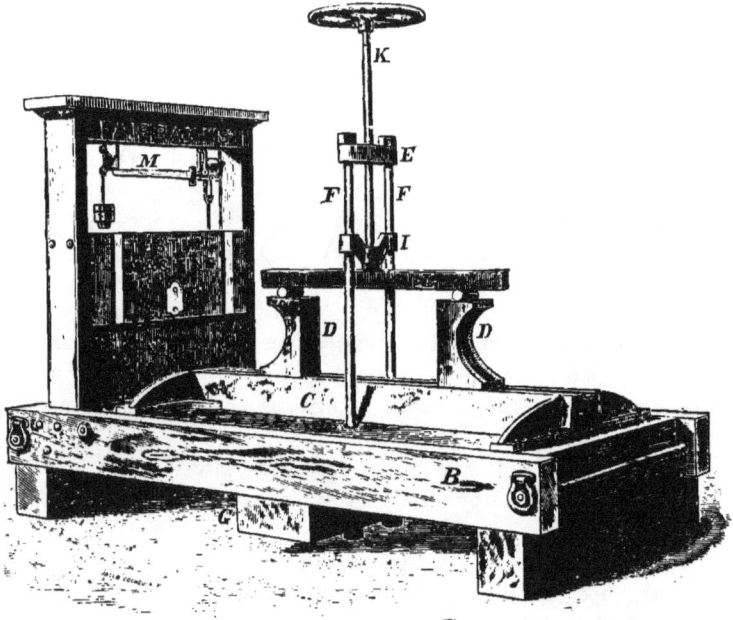

It consists of a Fairbanks Scale, on the platform of which rests a heavy cast iron beam, C, to which are fastened the supports, D, D, at the required distances apart. The pressure is applied by means of the hand wheel on the upper end of the screw, K, which screw passes through the nut, E, and terminates in the sliding cross-head, I. This cross-head serves both as a guide and as a pressure block. The test-piece, L, rests upon mandrels mounted upon the supports, D, D, at the required distance apart. The loads are weighed in the usual manner at M.

The instrument for measuring deflections is not shown in the cut; it consists of an accurately cut micrometer-screw of steel, having a pitch of 0.025 of an inch, working in a nut of the same material, mounted in a brass frame.. This instrument is supported by a rod of considerable rigidity and of sufficient length, which is secured to the beam, C, close to the tension-rods, F, F, in such a manner that the micrometer is directly over the cross-head, in the same vertical plane with the test-piece, and very near and parallel with the axis of the large screw, K. The micrometer-screw is provided with a head which is divided into 250 equal parts. Thus a rotary motion of one division produces an advance in the direction of the axis of the micrometer-screw of 0.0001 of an inch. A scale divided into fortieths of an inch is fastened to the frame of the instrument, in close proximity to the head, and parallel to the axis of the screw ; it serves to mark the starting point, and indicates the number of revolutions made in taking a measurement with the screw.

To insure accurate readings of the deflections, the principle of electric contact is here employed. Immediately underneath the micrometer-screw, on the cross-head, is placed an insulated metallic point, which is connected with one pole of one cell of a voltaic battery by means of an insulated conductor. The micrometerscrew is connected in a similar manner with the other pole. Thus, at the instant the screw touches the metallic point below it, connection is completed, and a current of electricity is established ; this fact is instantly indicated by the ringing of an electric bell placed in the circuit. By means of this instrument deflections of 0.0001 of an inch can be measured. The capacity of the testing machine is 7,000 pounds.

The test-pieces, size A, of the hot-rolled iron, and sizes A' and B' of the cold-rolled iron, were too heavy to be tested on the machine just described, the distance between the supports being insufficient to permit breaking them down under the maximum power of the machine.

The cold-rolled test-pieces, size B', were tested on the Fairbanks machine up to 7,000 pounds, on supports 52 inches apart, which was sufficient only to strain them slightly beyond the Elastic Limit, as will be seen on the curves. Sizes A and A' were also tested up to 7,000 pounds on this machine to determine the Modulus of Elasticity.

These large bars were finally broken down in the Tensile Testing Machine, to which a transverse testing attachment is fitted.

---

## DEFINITION OF TERMS.

(1.) The *Modulus of Elasticity* has been defined as the ratio of the distorting force to the amount of distortion, whether the latter be produced by extension or compression, so long as the Elastic Limit is not exceeded. For tension or compression, we have, therefore :

$$E = \frac{P}{l} \frac{L}{K} \dots\dots\dots\dots\dots\dots\dots\dots\dots(1)$$

The method of determining the Modulus of Elasticity by transverse stress is also based upon the above formula ; since a transverse stress necessarily produces a tensile and a compressive strain on opposite sides of the neutral surface of the bent bar ; this " neutral surface " is an unstrained section perpendicular to the distorting force at the point of its application, and passing through the centre of gravity of the cross-section of the unbent test-piece, or nearly so.*

By means of an analysis which would be out of place here, we find :

$$E = \frac{P L 3\dagger}{48 D I} \dots\dots\dots\dots\dots\dots\dots\dots\dots(2)$$

Where

E=the Modulus of Elasticity,
P=load,
D=the deflection due to P,
L= " length between the supports,
I = " Moment of Inertia, which is equal to $\frac{\pi r 4}{4}$, in round

bars, r being the radius.

---

* The neutral surface passes exactly through the centre of gravity of the cross-section of the bar, only when the material offers the same resistance to compression as to tension and is perfectly homogeneous.

†Wood's Resistance of Materials, pp. 105 and 113.

(2.) The *Elastic Limit* within which the above formula is applicable is reached when the rate of deflection changes, becoming greater, in proportion to, increase of load, than at the beginning of the test. This point is not marked on the strain-diagram.

The *practical* limit of elasticity as given in curves and records is generally considerably higher, and is taken at the point at which the permanent set becomes practically objectionable. This point may be found from the tables, but is more readily determined from the curves, where it is marked E.

In fixing the Elastic Limit for the test-pieces here reported upon, 0.05 inch has been assumed to be the greatest allowable set in a bar 1 inch square and 22 inches long.

(3.) The *Modulus of Rupture*, or the Modulus of Maximum Resistance, as determined by transverse stress, is defined to be the maximum strain upon a section of fibres one inch square most remote from the neutral surface, and on the side which first ruptures, if rupture occurs.* R, the Modulus of Maximum Resistance as given in the tables, is derived from the formula

$$R = \frac{P\,L\,r}{4\,I}\ \dots\dots\dots\dots\dots\dots\dots\dots(3)$$

in which P, L, I and $r$ are the same as in formula (2.)

(4.) *Transverse Resilience.* Elastic and Ultimate Resiliences have already been defined as being a measure of the capacity of a material to resist shock within the Elastic Limit, or up to the point of rupture, and their respective values are equal to the amounts of energy expended, or work performed, in springing a piece without injuring it or in producing fracture.

A true Modulus of Transverse Resilience cannot be obtained except by basing the calculations upon hypothesis, and making assumptions which experiments will not always justify. In the tables and in the following discussion, therefore, the Resiliences given are those of a standard bar of the metal tested, one inch square and 22 inches between the supports. This reduction of the Resiliences will answer quite well for purposes of comparison, and may also be

* Wood's Resistance of Materials, p. 156.

used as a basis for calculating the Resiliences of bars of any size of the same material. All the bars in this lot of test-pieces being too ductile to be broken, the Ultimate Resiliences could not be exactly determined, but are taken up to certain limits within those of rupture; *i. e.*, at the Elastic Limit and at deflections of two inches.

The Transverse Resilience at any given deflection is readily found from the appended strain-diagrams, plotted from the reduced figures. In these diagrams each inch of ordinate represents 500 pounds resistance, and each inch of abscissa measures one-half inch of deflection.

The formula is derived as follows :

Let $W$=the Resilience,

$P_m$=the mean stress applied,

$O$=the mean ordinate,

$S$=the abscissa of the point to which $W$ is to be calculated,

$D$=deflection at the same point,

$A$=the area, in square inches, included between $O$, $S$ and the curve.

According to definition,

$$W = P_m D \dots\dots\dots\dots\dots\dots\dots\dots\dots (4)$$

From the scale of the curve,

$$D = 0.5\,S$$

and

$$P_m = 500\,O$$

but

$$O = \frac{A}{S};$$

hence

$$P_m = 500\frac{A}{S};$$

substituting these values of $D$ and $P_m$ in formula (4), we have

$$W = 500\frac{A}{S} + 0.5\,S$$

$$= 250\,A \text{ inch-pounds of work,}$$

or

$$\tfrac{250}{12}\,A = 20.83\,A \text{ foot-pounds.}$$

The latter is used in the tables, as it is the more common unit of work.

(5) *Reduction of Results.* In order that the results of the tests of these bars of various sizes might be directly compared, they were all reduced to a common standard. That here adopted is the standard transverse test-piece of the Mechanical Laboratory of the Stevens Institute of Technology—a bar 1 inch square in section and 22 inches between supports.

The formula of reduction is derived as follows:

Let $P'$=the actual load on the round bars tested,

     $P''$=the load which would produce the same strain on a round bar, 1 inch in diameter,

     $P$=this load reduced to the standard,

     $d$=the diameter of the test-piece.

Supposing the length of the test-piece to be that of the standard (22 inches), we have:

$$P' : P'' : : d^3 : 1.$$

$$\therefore \quad P''=\frac{P}{d^3} \dots \dots \dots \dots \dots \dots (5)$$

Again, we have:

$$P : P'' : : 1 : 0.589,$$

0.589 being the strength of a cylindrical beam when that of a circumscribing square beam is taken as unity.*

$$\therefore P''=0.589\ P,$$

and by substituting above

$$P=\frac{P'}{0.589 d^3}$$

For any other length, L, of test-piece than the standard (22 inches), we have:

$$P=\frac{L}{22} \times \frac{P'}{0.589 d^3}$$

The *Reduction of Deflections* of bars of lengths, other than 22 inches, was effected by the aid of the proportion actual Deflection : reduced Deflection : : L : 22, which gives Reduced Deflection= $\frac{22}{L} \times$ Actual Deflection, L being the length, in inches, of the bar.†

*Plotting of the Curves.* From the results obtained from formula (6), the accompanying strain-diagrams were plotted.

---

  * Wood's Resistance of Materials, p. 182.

  † This formula of reduction is only approximately true.

Each inch on the horizontal scale corresponds to 0.5 inch deflection, and each inch on the vertical scale represents 500 pounds of stress.

The *Curves of Sets* which are drawn with the strain-diagrams show the amounts of permanent distortion for all loads. The horizontal distance between the initial ordinate and any point in the curve of sets measures the set, or permanent distortion produced by the load indicated by the height of that point, and the horizontal distance between the curve of sets and the strain-diagram measures the amount of recoil or spring when the load is entirely removed.


## DETAILS OF TESTS.

Nos. 1106A, 1107A and 1108A were all of hot-rolled iron, 2.54 inches in diameter—rough; they were tested on the transverse attach· ment to the Tension Testing-Machine with a distance between the supports of 22 inches. The loads were applied in increments of 1,000 pounds and the deflections were measured for every load, and are recorded in the tables. The sets were observed for each addition of 2,000 pounds.

**No. 1106A.** This test-piece deflected with a fair degree of regularity within the Elastic Limit, passing that limit under a load of 14,500 pounds, equivalent to 1,500 pounds on the standard bar; it then deflected more and more rapidly and with some irregularity (see curve of this number, Plate X), until the maximum load, 25,000 pounds,—equivalent to 2,590 pounds on the standard bar— was reached with a deflection of 2.90 inches. Here the test was discontinued.

The Modulus of Elasticity is 25,422,000, the Elastic Resilience is 10.31, and the Ultimate—at 2 inches deflection—is 339.74 foot-pounds. The Modulus of Maximum Resistance was 85,500. This bar showed no signs of rupture after the test.

**No. 1107A.** This bar deflected quite regularly within the Elastic Limit, which it passed at a load of 14,000 pounds—1,450 on the standard. It then deflected very regularly and reached its maximum load, 26,500 pounds—equivalent to 2,745 pounds on the standard test-piece—at a deflection of 2.9 inches. The resistance then began to decrease ; with 3.6 inches deflection, it bore but 25,200—or 2,611 pounds of load reduced to standard.

The Modulus of Elasticity is 25,659,000 ; that of Maximum Resistance is 90,600 pounds. The Elastic Resilience is 7.55, and at 2 inches deflection the Resilience amounts to 338.90 foot-pounds. No signs of rupture appeared after testing.

**No. 1108A.** This test-piece deflected very regularly within the Elastic Limit. It passed the limit under a load of 14,500 pounds —1,450 pounds standard—after which it deflected rapidly and regularly and reached its Maximum Resistance, 26,000 pounds— equivalent to 2,694 pounds on the standard test-piece—with a deflection of 2.8 inches. With a deflection of 3.39 inches it still carried 25,500 pounds—2,642 pounds on the standard bar ; but here it suddenly yielded, carrying but 14,275 pounds—equivalent to 1,470 pounds of reduced load—at a deflection of 3.45 inches. The test was here discontinued.

The Modulus of Elasticity is 26,146,000 ; that of Maximum Resistance is 88,890 pounds. The Elastic Resilience is 8.44, and the total Resilience at 2 inches deflection is 342.66 foot-pounds. This bar showed indications of rupture at the close of test.

Nos. 1109A′, 1110A′ and 1111A′ were of cold-rolled iron, $2_{\frac{7}{16}}$ inches in diameter ; they were tested like the preceding, except that the distance between supports was 33 inches.

**No. 1109A′.** This bar deflected regularly within the Elastic Limit, which it passed under a load of 18,000 pounds, equivalent to 2,165 on the reduced scale ; it then deflected more and more rapidly and with fair regularity until it reached its maximum load, 23,000 pounds—equivalent to 4,040 pounds standard—with a deflection of 2.30 inches. At a deflection of 2.38 inches the bar sustained 22,550 pounds—equivalent to 3,965 pounds on the standard bar—having there suddenly weakened, (see curve No. 1109A′, Plate X).

The Modulus of Elasticity of this bar is 29,998,000 ; that of the Maximum Resistance is 133,480 pounds. The Elastic Resilience is 42.29 foot-pounds, and at a deflection of 2 inches the Resilience is 575.12 foot-pounds. No signs of rupture could be detected at the end of the test.

**No. 1110A′** behaved very similarly to the preceding test-piece while under stress. It passed its Elastic Limit under a load of 18,000 pounds—equivalent to 3,165 pounds on the standard bar. It then deflected quite regularly and more and more rapidly, and reached its maximum load, 23,200 pounds—equivalent to 4,080 pounds standard—with a deflection of 2.27 inches; it then lost strength, sustaining only 22,290 pounds—equivalent to 3,920 pounds of reduced load—with a deflection of 2.38 inches.

The Modulus of Elasticity is 28,195,000; that of Maximum Resistance is 134,640 pounds. The Elastic Resilience is the same as that of the preceding one—42.19 foot-pounds; at 2 inches deflection, the total Resilience is 573.49 foot-pounds. No visible rupture was found after the test.

**No. 1111A′.** This test-piece deflected even more regularly within the Elastic Limit than did its two companion specimens, and it passed the latter point under the same load—18,000 pounds, or 3,165 pounds on the reduced scale. After passing the Elastic Limit, it deflected rapidly, but not quite as regularly as did the two preceding test-pieces. It reached the Maximum Stress, 23,200 pounds, which is equivalent to 4,080 pounds of reduced load, at a deflection of 1.99 inches. The Modulus of Elasticity is 30,843,000 ; that of Maximum Resistance is the same as that of the preceding test-piece, 134,640 pounds. The Elastic Resilience is 4.670 foot-pounds, which is slightly less than that of the preceding ; at a deflection of 2 inches, the Resilience is 579.91 foot-pounds. No signs of rupture were visible after the test.

**No. 1128A′** was cold-rolled and annealed iron, off the same bar as the other cold-rolled bars, marked A′. The distance between the supports was 22 inches, and the load was applied in increments of 1,000 pounds. This test-piece passed its Elastic Limit under a load of 15,380 pounds, which is equivalent to 1,800 on a standard bar. It then deflected with considerable irregularity (see curve No. 1128A′, Plate X), then still more regularly, and finally carried the maximum load, 24,500 pounds—equivalent to 2,870 pounds of reduced load—at a deflection of 2.56 inches, after which it weakened very rapidly, sustaining only 18,300 pounds, or 2,145 pounds standard, at a deflection of 2.68 inches.

The Modulus of Elasticity is 30,539,000 ; that of Maximum Resistance is 94,790 pounds. The Elastic Resilience is 11.25 foot-pounds, and at 2 inches the total Resilience is 394.31 foot-pounds. No signs of rupture appeared after the test.

The test-pieces Nos. 1106B, 1107B and 1108B were hot-rolled iron, 2.08 inches in diameter. These bars were all tested on our Fairbanks transverse testing-machine, with a distance of 44 inches between the supports. The loads were applied in increments of 200 pounds, and the deflections recorded for every load, as seen in the tables, where the actual as well as the reduced figures are given.

No. 1106B passed its Elastic Limit under a load of 3,630 pounds, equivalent to 1,375 pounds on the standard bar. It then deflected rapidly, but with some irregularity, until it reached its maximum— 5,750 pounds, equivalent to 2,170 pounds standard—after having deflected 5.5 inches—equivalent to 2.75 inches with the standard bar. At a deflection of 6 inches—equivalent to 3 inches deflection of the standard test-piece—it still sustained the same load. Here the test was discontinued.

The Modulus of Elasticity is 28,197,000, while that of Maximum Resistance is 71,590 pounds. The Elastic Resilience is 9.74 foot-pounds, and at 2 inches of deflection the total Resilience is 282.49 foot-pounds.

No. 1107B deflected somewhat irregularly within the Elastic Limit, which it passed under a load of 3,580 pounds—equivalent to 1,350 pounds on the standard bar. It then deflected rapidly and very regularly until the load became 4,000 pounds, when the bar suddenly *stiffened*, as is best shown by the offset in curve No. 1107B, Plate XI. It gave a Maximum Resistance of 5,950 pounds, with a deflection of 8 inches; this on a standard bar would be equivalent to 2,245 pounds and a deflection of 4 inches. The test was then discontinued.

The Modulus of Elasticity is 27,592,000 ; that of Maximum Resistance is 74,080 pounds. The Elastic Resilience is 8.44 foot-pounds, and the total Resilience at a deflection of 2 inches is 286.00 foot-pounds. No signs of rupture were discovered after the test.

No. 1108B passed its Elastic Limit under the same load as did the preceding test-piece, 3,580 pounds, equivalent to 1,350 pounds

on the standard bar; it then deflected rapidly and somewhat irregularly, (see curve No. 1108B, Plate XI,) until it reached a Maximum Resistance of 5,880 pounds at a deflection of 7 inches, which
on the standard bar would be equivalent to a load of 2,220 pounds
and a deflection of 3.5 inches.

The Modulus of Elasticity is 28,404,000; the Modulus of Maximum Resistance is 73,210 pounds. The Elastic Resilience of this
test-piece is 8.59 foot-pounds, and the Resilience, at a deflection of
2 inches, is 282.49 foot-pounds. No signs of rupture could be
detected on the surface of the bar.

Bars Nos. 1109B', 1110B' and 1111B' were of cold-rolled iron
shafting, 2 inches in diameter. These test-pieces were tested upon
the Fairbanks transverse testing machine up to 7,000 pounds—the
full capacity of the machine; the distance between the supports
was made 52 inches—the full length of the specimen. The load
was applied in increments of 250 pounds, and the deflections were
recorded as usual.

No. 1109B' was first tested with a distance of 44 inches between
the supports, and increments of loads of 200 pounds up to 7,000
pounds; the supports were then put 52 inches apart, and the loads
again applied and deflections measured as before. Thus these three
test-pieces were strained considerably beyond their Elastic Limits,
but not nearly to their Maximum Resistance. They were finally
tested on the altered tension machine; the details of these tests are
given in the tables. Curves Nos. 1109B', 1110B' and 1111B',
Plate XI, exhibit graphically the behavior of the specimens in both
tests. The portions a b, of the curves, are plotted from the results
of the initial tests, and the dotted continuation, b c, is supposed to
be the curve which would have been given if the tests could have
been carried through without intermission, thus making smooth
curves a b c, whereas the actual curves are a b b', which have a
sudden rise at b, showing a considerable elevation of ultimate
strength caused by their having been previously strained beyond
the Elastic Limits. The curves a' b b' were obtained from the
final tests. In the tables of comparative results the probable values
deduced from the curves a b c will only be given; but in the following discussion the results of both will be stated and compared.

No. 1109B,' as represented by curve *a b c*, passed its Elastic Limit under the standardized loads,* 2,950 pounds, after which the deflections for equal increments of load increased gradually for some distance, then more and more rapidly, until the probable maximum resistance of 4,200 pounds was reached. As represented by curve *a' b b'*, on the second test, it passed its Elastic Limit under a load of 3,700 pounds, which is 750 pounds more than before, and it reached its maximum strength at 4,350 pounds and a deflection of 2.35 inches. The Elastic Resiliences are 45.47 and 40.08 foot-pounds, respectively; whereas the Resiliences at 2 inches deflection are, from curve *a b c*, 560.54, and from *a' b b'* 628.03 foot-pounds.

The Modulus of Elasticity of this bar is 27,896,000. The Modulus of the probable Maximum Resistance is 136,000, while that obtained from final test is 143,500 pounds, showing an increase of 7,500 pounds.

No. 1110B', as represented by the curve *a b c*, passed its Elastic Limit under a load of 3,050 pounds, after which the deflections increased until the probable maximum was reached, at about 4,120 pounds. On the curve *a' b b'* the Elastic Limit is noted at a load of 3,650 pounds, which is an increase over that of the initial test, of 600 pounds. It reached its Maximum Strength, 4,300 pounds, with a deflection of 2.41 inches. The Elastic Resiliences are, respectively, 54.00 and 36.48 foot-pounds; while the Resiliences at 2 inches are, respectively, 545.12 and 622.61 foot-pounds.

The Modulus of Elasticity is 26,452,000; the Moduli of the probable Maximum Resistances, and of that obtained from the final test are, respectively, 135,200 and 142,100; the latter being the greater by 6,900 pounds.

No. 1111B' passes the Elastic Limit on the curve *a b*, at a load of 3,050 pounds, after which the curve runs very much like the preceding, reaching the probable Maximum of Resistance of 4,100. On the curve *a' b b'*, the Elastic Limit is passed under a load of 3,850 pounds—700 pounds higher than on the partial test. The Maximum Resistance at the final test was 4,310 pounds—210 pounds more than at the previous test. The deflection for the maximum

*For actual loads, consult the tables.

load was 2.11 inches, after which it suddenly lost strength, resisting only 3,862 pounds at a deflection of 2.27 inches.

The Elastic Resiliences are, respectively, 52.87 and 48.12 foot-pounds. The Resiliences at 2 inches deflection are, respectively, 550.75 and 644.69 foot-pounds. The Modulus of Elasticity is 28,256,000, and the Moduli of Maximum Resistances are, respectively, 142,100 and 135,000 foot-pounds, the probable Modulus being 7,100 pounds below that obtained at the second test.

**No. 1128B'**, of cold-rolled iron and annealed, was 2 inches in diameter, and was tested at a distance of 38 inches between the supports. The load was applied by adding 500 pounds at a time up to 3,000 pounds, then by increments of 250 pounds. The deflections were noted as before. This test-piece passed its Elastic Limit under a load of 5,030 pounds, equivalent to 1,850 on the standard bar. It then deflected regularly, and reached its Maximum Resistance, 7,000 pounds, at a deflection of 4.35 inches, which is equivalent to 2,570 pounds, and to a deflection of 2.52 inches for a standard bar.

The Modulus of Elasticity is 27,463,000; that of Maximum Resistance is 84,670 pounds. The Elastic Resilience is 15.41 foot-pounds. At a deflection of 2 inches the Resilience is 354.74 foot-pounds. No sign of rupture appeared after the test was completed.

Nos. 1106C, 1107C and 1108C of hot-rolled iron; Nos. 1109C' 1110C' and 1111C' of cold-rolled iron, and No. 1128C' (C is 1.36 and C' 1$\frac{5}{16}$ inches in diameter) were tested with a distance of 30 inches between the supports. The stress was applied by increments of 100 pounds, and the deflections noted for every load; the sets were measured for each additional 300 pounds of stress.

**No. 1106C** passed its Elastic Limit under a load of 1,410 pounds —equivalent to 1,300 pounds on the standard bar—after which it bent rapidly and pretty regularly until it reached its Maximum Resistance, 2,170 pounds, with a deflection of 2.58 inches when reduced to the standard.

The Modulus of Elasticity of this bar is 28,143,000; that of Maximum Resistance is 65,900 pounds. The Elastic Resilience is 7.52 foot-pounds, and at a deflection of 2 inches of Resilience amounts to 265.56 foot-pounds. After the deflection of 3.5 inches —2.58 on the standard—was reached, the test was discontinued.

No. 1107C passed its Elastic Limit under a load of 1,448 pou: ds, equivalent to 1,330 pounds of standardized load—after which it deflected rapidly but less regularly than the preceding bar, until it reached its Maximum Resistance of 2,440 pounds with a deflection of 2.6 inches, which, when reduced to the standard, is equivalent to 1,959 pounds of load and 1.91 inches of deflection. The test was continued until the deflection was 4 inches, at which the load was 2,135 pounds—equivalent to 2.93 inches deflection, and 1,964 pounds of load per standard bar.

The Modulus of Elasticity is 28,624,000, and that of Maximum Resistance is 64,990 pounds. The Elastic Resilience is 8.64 foot-pounds, and at a deflection of 2 inches the total Resilience is 270.16 foot-pounds.

No. 1108C. This bar deflected more regularly within the Elastic Limit than did its two companion test-pieces ; it passed the limit under a load of 1,468 pounds—equivalent to 1,350 pounds per standard. It then deflected rapidly and uniformly, reaching its Maximum Resistance, 2,125 pounds, with a deflection of 3 inches, which standardized, is equivalent to 1,956 pounds of stress and a deflection of 2.19 inches. The resistance then decreased, and at a deflection of 3.5 inches it was only 2,060 pounds, or in reduced figures, 2.56 inches deflection and 1,896 pounds of stress. Here the test was discontinued.

The Modulus of Elasticity is the same as that of the preceding bar—28,624,000—and the Modulus of Maximum Resistance is 64,540 pounds. The Elastic Resilience is 9.56 foot-pounds ; at a deflection of 2 inches the Resilience is 269.54 foot-pounds.

The Maximum Resistances, and consequently the Resiliences, at 2 inches deflection for the three preceding test-pieces are exceedingly low, compared with the other sizes. The maximum load was also reached with a less deflection than with the other bars of untreated iron.

No. 1109C', of cold-rolled iron, deflected very regularly within the Elastic Limit, which it passed under a load of 2,392 pounds, equivalent to 2,450 pounds on the standard bar ; it then deflected more and more rapidly and quite uniformly, until the Maximum Resistance, 3,400 pounds, was reached at a deflection of 3 inches— equivalent on the standard bar to 3,482 pounds of stress and 2.2

inches of deflection. With a deflection of 3.5 inches the load reduced to 3,355 pounds—equivalent to 2.57 inches of deflection and 3,436 pounds of stress on the standard.

The Modulus of Elasticity is 26,625,000, and that of Maximum Resistance is 114,890 pounds. The Elastic Resilience is 26.45 foot-pounds, and at a deflection of 2 inches the Resilience is 494.72 foot-pounds.

**No.** 1110C' passed its Elastic Limit under a load of 2,600 pounds, and then deflected more and more rapidly and with regularity, reaching its maximum load, 3,470 pounds, at a deflection of 3 inches—equivalent to 3,553 pounds on the standard bar and a deflection of 2.20 inches. The test was discontinued when the deflection had reached 3.5 inches, the resistance having decreased to 3,290 pounds—equivalent to 2.57 inches of deflection and a stress of 3,370 pounds when reduced to the standard.

The Modulus of Elasticity is 25,655,000, and that of Maximum Resistance is 117,620 pounds. The Elastic Resilience is 32.97 foot-pounds, and at 2 inches of deflection the Resilience is 500.34 foot-pounds.

**No.** 1111C' was more irregular in its deflection within the Elastic Limit than its two companion specimens. It passed its Elastic Limit under a stress of 2,550 pounds, equivalent to 2,490 pounds on the standard bar; it then deflected very similarly to the two preceding bars, reaching its Maximum Resistance, 3,500 pounds, with a deflection of 3.5 inches—equivalent to 3,584 pounds of stress and 2.56 inches of deflection per reduced scale.

The Modulus of Elasticity is 27,671,000, and that of Maximum Resistance is 118,390 pounds. The Elastic Resilience is 34.76 foot-pounds, and at a deflection of 2 inches the Resilience is 503.46 foot-pounds. The Elastic Limit and the Maximum Resistance, and consequently the Elastic Resilience as well as the Resilience at 2 inches deflection for size C', will be found much lower than any of the other cold-rolled bars; this may be due partly to the fact that the hot-rolled bars of size C were not as strong as the other sizes; but it is perhaps due still more to the fact that the percentage of reduction by cold-rolling is least in size C'. The percentages of reduction by cold-rolling run as follows: A, 4 per cent.; B, 3.85 per cent.; C, 3.48 per cent.; D, 3.84 per cent.; and E, 6 per cent.

No. 1128C' was of cold-rolled and annealed iron, and was tested in all respects like the preceding bars. It deflected very uniformly within the Elastic Limit, which it passed under a stress of 1,844 pounds—equivalent to 1,900 pounds on the standard bar ; it then deflected more and more rapidly, and quite regularly, and reached its Maximum Resistance at a deflection of 4 inches, which, reduced to the standard, is equivalent to a load of 2,500 pounds and a deflection of 2.93 inches.

The Modulus of Elasticity is 27,188,000, and that of Maximum Resistance is 82,620 pounds. The Elastic Resilience is 19.00 foot-pounds, and at a deflection of 2 inches the Resilience is 352.90 foot-pounds.

It is remarkable that although the hot-rolled and the cold-rolled bars of size C are inferior, the cold-rolled and annealed bar of the same size should be superior in many of its qualities to the annealed bars of the other sizes.

Nos. 1106D, 1107D and 1108D were of untreated iron ; the diameter of each was 1.04 inches, and the distance between the supports was made 22 inches. The stress was applied by increments of 40 pounds, and the deflections measured as before. The sets were measured for every 200 pounds, and the tests were tabulated as before.

No. 1106D passed its Elastic Limit under a load of 1,140 pounds —equivalent to 1,720 pounds on the standard test-piece ; it then deflected very irregularly, (see curve No. 1106D, Plate XIII) showing five distinct and almost complete cessations of molecular "flow" before reaching its Maximum Resistance—1,620 pounds— equivalent to 2,455 pounds on the reduced scale, with a deflection of 4 inches.

The Modulus of Elasticity is 25,782,000, and that of Maximum Resistance is 80,690 pounds. The Elastic Resilience is 14.37 foot-pounds, and at a deflection of 2 inches the total Resilience is 317.92 foot-pounds.

The sudden strengthening of this and of the next two bars at various stages of the test, so plainly exhibited by the curves, is not due to the method of testing, since these bars were not allowed to rest at any point, but were tested as continuously as all the others, which show no such peculiarities ; the same must therefore be some peculiarity of structure in the material itself.

**No. 1107D.** This bar deflected quite uniformly within the Elastic Limit, which it passed under a stress of 1,160 pounds—equivalent to 1,750 pounds per reduced scale. It then deflected rapidly and almost as irregularly as the preceding test-piece, showing similar abrupt elevations in the strain-diagram due to the same cause. It reached its Maximum Resistance, 1,650 pounds—equivalent to 2,490 pounds on the standard bar—at a deflection of 4 inches. Here the test was discontinued.

The Modulus of Elasticity is 27,233,000, and that of Maximum Resistance is 82,180. The Elastic Resilience is 14.58 foot-pounds, and at a deflection of 2 inches the Resilience is 322.09 foot-pounds.

**No. 1108D** deflected very regularly within the Elastic Limit, which it passed under the same load as did the preceding test-piece, viz., 1,160 pounds—equivalent to 1,750 pounds on the standard bar. It then deflected somewhat more uniformly than did the two preceding specimens, but still very irregularly, until it reached its Maximum Resistance, 1,700 pounds—equivalent to 2,565 pounds on the reduced scale—with a deflection of 4 inches.

The Modulus of Elasticity of this test-piece is 25,782,000, and that of Maximum Resistance is 84,670 pounds. The Elastic Resilience is 14.52 foot-pounds, and at a deflection of 2 inches the Resilience is 326.56 foot-pounds.

Nos. 1109D', 1110D' and 1111D' were of cold-rolled iron, 1 inch in diameter, and were tested under exactly the same conditions as were the three bars of which the tests have just been described.

**No. 1109D'** deflected very regularly within its Elastic Limit, which it passed under a load of 1,780 pounds—equivalent to 3,000 pounds per reduced scale. It then deflected more and more rapidly and pretty regularly, excepting that a rapid increase of resistance was observed just before reaching the maximum load. The latter, 2,450 pounds—equivalent to 4,150 pounds of reduced load—was reached with a deflection of 2.13 inches, after which the strength decreased gradually, until at a deflection of 4 inches it carried only 2,370 pounds—equivalent to 4,024 pounds on the standard bar.

The Modulus of Elasticity is 27,094,000 ; that of Maximum Resistance is 136,740 pounds. The Resilience at the Elastic Limit is 43.74 foot-pounds, and that at 2 inches deflection is 571.15 foot-pounds.

No. 1110D' deflected quite regularly within the Elastic Limit, which it passed under a stress of 1,800 pounds—equivalent to 3,056 pounds on the reduced scale—after which it deflected more and more rapidly, although not very regularly, showing some of the peculiarities of the untreated bars ; it reached its Maximum Resistance, 2,440 pounds—equivalent to 4,143 pounds on the standard bar—with a deflection of 2 inches. After this the resistance decreased, bearing only 2,350 pounds—equivalent to 3,990 pounds on the standard—with a deflection of 4 inches.

The Modulus of Elasticity is 27,163,000, and that of Maximum Resistance 136,690 pounds. The Elastic Resilience is 44.47 foot-pounds, and at a deflection of 2 inches the Resilience amounts to 582.19 foot-pounds.

No. 1111D' deflected with considerable regularity within the Elastic Limit. It passed that point under a load of 1,800 pounds —equivalent to 3,056 pounds on the standard bar. It then deflected more and more rapidly, and reached its Maximum Resistance, 2,400 pounds—equivalent to 4,076 pounds on the reduced scale—with a deflection of 1.71 inches. After passing its maximum it very gradually decreased in resistance, until, with a deflection of 4 inches, it had reduced to 2,335 pounds—equivalent to 3,977 pounds of reduced load.

The Modulus of Elasticity of this bar is 27,094,000, and the Modulus of Maximum Resistance is 133,430 pounds. The Elastic Resilience is 47.22 foot-pounds, and at a deflection of 2 inches the Resilience is 576.46 foot-pounds.

Nos. 1106E, 1107E and 1108E were test-pieces of hot-rolled iron, 0.665 inch in diameter. They were tested with a distance between the supports of 22 inches. The stress was applied in increments of 20 pounds, and the deflection measured and recorded as before. The sets were observed and noted for every 200 pounds of load.

No. 1106E passed its Elastic Limit under a stress of 303 pounds —equivalent to 1,750 pounds on the standard bar—after which it deflected very rapidly until its maximum, 380 pounds—equivalent to 2,194 pounds per reduced scale—was reached, it having deflected 3.25 inches. Here the test was discontinued.

The Modulus of Elasticity is 28,528,000, and that of Maximum Resistance is 72,390 pounds. The Elastic Resilience is 21.88 foot-pounds ; the Resilience at 2 inches deflection is 298.13 foot-pounds.

**No. 1107 E.** This test-piece passed its Elastic Limit under a stress of 303 pounds—equivalent to 1,750 pounds on the standard bar. It then bent very rapidly and quite regularly until the Maximum Resistance, 380 pounds—equivalent to 2,194 pounds of reduced load—was reached with a deflection of 3.85 inches. Here the test was stopped.

The Modulus of Elasticity is 28,705,000, and that of Maximum Resistance is 72,390 pounds. The Resilience at the Elastic Limit is 21.88 foot-pounds, and that at 2 inches deflection is 292.71 foot-pounds.

**No. 1108E** passed its Elastic Limit under the same load as did the preceding test-piece, and deflected in a similar manner, reaching its Maximum Resistance, 380 pounds—equivalent to 2,194 pounds on the standard bar—with a deflection of 3.5 inches, at which point the test was discontinued.

The Modulus of Elasticity is 28,343,000, and the Modulus of Maximum Resistance is the same as in the two preceding cases—72,390 pounds. The Elastic Resilience is 20.41 foot-pounds, and at a deflection of 2 inches the Resilience is 294.59 foot-pounds.

Nos. 1109E′, 1110E′ and 1111E′ were test-pieces of cold-rolled iron, ¾ inch in diameter.

**No. 1109E′** passed its Elastic Limit under a load of 410 pounds —equivalent to 2,850 pounds of reduced stress. Its deflections then very gradually, but not very rapidly, increased, until the load of 570 pounds—equivalent to 3,964 pounds—was reached; it then gave way quite rapidly, its resistance being a maximum at a deflection of 1.53 inches. The maximum load was 580 pounds— equivalent to 4,033 pounds per reduced scale.

The Modulus of Elasticity is 27,443,000, and that of Maximum Resistance is 133,090 pounds. The Elastic Resilience is 55.80 foot-pounds, and at a deflection of 2 inches the Resilience measured 545.62 foot-pounds.

No. 1110E' had the same Elastic Limit as had the preceding test-piece; it then deflected similarly, but more regularly, giving the same Maximum Resistance, which it reached with a deflection of 2 inches.

The Modulus of Elasticity is 27,094,000 ; the Modulus of Maximum Resistance is the same as in the preceding case. The Elastic Resilience is 56.99 foot-pounds ; the Resilience, at a deflection of 2 inches, is 540.21 foot-pounds.

No. 1111E' passed the Elastic Limit under the same stress as did its companion test-pieces, after which it, however, deflected more rapidly and more uniformly, but reached the same Maximum Resistance at a deflection of 2 inches.

The Moduli of Elasticity and of Maximum Resistance were the same as those of the preceding test-piece, viz., 27,094,000 and 133,-090 pounds respectively. The Elastic Resilience is 58.18 foot-pounds, and at a deflection of 2 inches the Resilience is 529.27 foot-pounds.

No. 1128E', of cold-rolled and annealed iron, was tested in the same manner as were the preceding test-pieces, of size E ; it passed its Elastic Limit under a load of 260 pounds—equivalent to 1,808 pounds on the standard bar—after which the deflection increased more and more rapidly until the Maximum Resistance, 342 pounds, —equivalent to 2,378 pounds on the reduced scale—with a deflection of 3.5 inches it still carried the same load, after which the resistance decreased.

The Moduli of Elasticity and of Maximum Resistance are respectively 26,687,000 and 78,480 pounds. The Elastic Resilience is 24.75 foot-pounds, and at a deflection of 2 inches the Resilience amounts to 333.96 foot-pounds.

# RESUMÉ.

## I.—HOT–ROLLED, OR UNTREATED IRON.

In reviewing the results which have been discussed in the preceding pages, and comparing the tables and curves, we find :

(1.) That the average strength of the hot-rolled iron, when reduced to the standard bar, is about 2,300 pounds, which gives an average Modulus of Resistance of 76,000 pounds. From Table I of Comparative Results, we see that the *largest* bar has the *greatest* Modulus of Transverse Resistance, and that it does not decrease regularly with the diameter of the test-piece—size D ($1\frac{1}{16}$ inches diameter), having a very high Modulus of Resistance, only 5,800 pounds less than that of size A. With the exception of size D, the decrease of the Modulus of Resistance with the diminution of the diameters of the bar are very marked. In tension, the Modulus *increased* as the diameters *decreased*, but not in so marked a degree. The curious fact that the capacity of size D to resist transverse stress is above the average, while it falls below the average in resisting tensile strength is as probably due to differences in the iron as to differences in distribution of resisting power in the section of the bar. With size C ($1\frac{3}{4}$ inches diameter) the reverse is the case.

(2.) That the average Elastic Limit is 1,550 pounds, and that this approaches the Maximum Resistance as the diameters of the test-pieces decrease, as is best observed from the curves CC and C'C', Plate XV, which approach each other more closely as they pass from A to E.

(3.) That the Resiliences vary with the Elastic Limits and with the Maximum Resistances of the test-pieces; the Elastic Resiliences being greatest with the smallest bars and the Resiliences at deflections of 2 inches increasing with the sizes of the test-pieces ; but without any regularity in either case.

(4.) That the Moduli of Elasticity vary between 25,000,000 and 29,000,000, and that they are independent of the diameters of the test-pieces.

## II.—COLD-ROLLED IRON.

From a study of all the records of tests of cold-rolled iron which were individually discussed in the preceding pages, we learn:

(1.) That the Maximum Resistance of a cold-rolled bar of the standard size is 8,980 pounds—1,665 pounds greater than that of the hot-rolled iron—giving a Modulus of 131,000 pounds—55,100 pounds greater than that of the latter. Except with size C ($1\frac{5}{16}$ inches diam.), there is no marked difference in the Moduli of Resistances for the different diameters, as is shown by the curve AA, Plate XV. This may be partly due to the fact that the smaller sizes, except C, were reduced more by cold-rolling than were the larger ones.

(2.) That the average Elastic Limit is 3,040 pounds—1,490 pounds greater, or nearly double, that of the hot-rolled iron. The fact observed in the untreated iron that the Elastic Limit approaches the ultimate strength as the diameters decrease, is not true for the cold-rolled iron; but it appears nevertheless to be more dependent upon the ultimate strength than is the case in tension.

(3.) That the Resiliences, Elastic and Ultimate, are independent of the diameters of the test-pieces. The average Elastic Resilience is 45.2 foot-pounds, while that of hot-rolled iron is only 13.82 foot-pounds.

(4.) That the Moduli of Elasticity range between 25,000,000 and 30,000,000, and are independent of the diameters of the bars tested.

## III.—COLD-ROLLED AND ANNEALED IRON.

The results of these tests also show:

That by annealing the cold-rolled iron it loses all its characteristic properties in a very marked degree, behaving under stress more like untreated than like cold-rolled iron.

The Elastic Limit and Maximum Resistance are only slightly higher than those of the untreated iron. This difference, however, is more marked under transverse stress than in tension. The tensile resistance of the annealed iron is but very little greater than that of the untreated iron.

# GENERAL CONCLUSIONS.

From a study of the accompanying strain-diagrams and the appended tables, as well as from what has previously been stated, the following general conclusions may be readily drawn:

(1.) The process of cold-rolling produces a very marked change in the physical properties of the iron thus treated:

(a.) It increases the tenacity from 25 to 40 per cent., and the resistance to transverse stress from 50 to 80 per cent.

(b.) It elevates the Elastic Limits under both tensile and transverse stresses, from 80 to 125 per cent.

(c.) The Modulus of Elastic Resilience is elevated from 300 to 400 per cent. The Elastic Resilience to transverse stress is augmented from 150 to 425 per cent.

(2.) Cold-rolling also improves the metal in other respects:

(a.) It gives the iron a smooth, bright surface, absolutely free from the scale of black oxide unavoidably left when hot-rolled.

(b.) It is made exactly to gauge, and for many purposes requires no further preparation.

(c.) In working the metal the wear and tear of the tools are less than with hot-rolled iron, thus saving labor and expense in fitting.

(d.) The cold-rolled iron resists stresses much more uniformly than does the untreated metal. Irregularities of resistance exhibited by the latter do not appear in the former; this is more particularly true for transverse stress, as is shown by the smoothness of the strain-diagrams produced by the cold-rolled bars.

(e.) This treatment of iron produces a very important improvement in uniformity of structure, the cold-rolled iron excelling common iron in its uniformity of density from surface to centre, as well as in its uniformity of strength from outside to the middle of the bar.

(3.) This great increase of strength, stiffness, Elasticity and Resilience is obtained at the expense of some ductility, which diminishes as the tenacity increases. The Modulus of Ultimate Resilience of the cold-rolled iron is, however, above 50 per cent. of that of the untreated iron.

Cold-rolled iron thus greatly excels common iron in all cases where the metal is to sustain maximum loads without permanent set or distortion.

(4.) We conclude, that as a material of construction, cold-rolled iron has many peculiar advantages; that it is suitable for all constructions not exposed to high temperatures; that it is especially suitable for all purposes demanding a high Elastic Limit and great shock resisting power without permanent distortion; that the process improves the metal throughout—its benefit, as has been seen, reaching the centre of the bar, and rendering the whole much more homogeneous and uniform than common iron— and that in many cases it may prove superior even to some steels as a material of construction.

The tables of summaries will be found particularly convenient as exhibiting most concisely all data obtained. The table of working and breaking loads, Table L, page 88, will be of especial value for many cases of practical application.

Very Respectfully,

R. H. THURSTON.

# TABLES

# FINAL SUMMARIES OF COMPARABLE RESULTS.

## LOT No. 1.

### A—COLD-ROLLED IRON.

All specimens of this Lot were tested without previous preparation in the lathe.

| Lab. No. | Orig-inal Mark. | Diam-eter. inches. | Elastic Limit. | Modulus of Rupture per square inch of | | Modulus of Resil-ience. | Per cent. of Exten-sion. | Per ct. of Re-duction of area at Frac-ture. |
|---|---|---|---|---|---|---|---|---|
| | | | | Original Section. | Fractured Section. | | | |
| 1133 | 1-66 | $2\frac{7}{16}$ | 59,450 | 64,800 | 67,400 | 600 | 1.15 | 8.86 |
| 1134 | 2-66 | " | 59,450 | 66,500 | 72,700 | 1,672 | 2.75 | 8.62 |
| 1135 | 3-66 | " | 62,000 | 67,000 | 75,300 | 2,770 | 4.35 | 10.95 |
| Av'rage. | ...... | ...... | 60,300 | 66,100 | 71,800 | 1,681 | 2.75 | 7.81 |
| 1140* | 1-66 | 2 | 57,500 | 66,400 | 91,800 | 3,107 | 5.60 | 27.74 |
| 1141* | 2-66 | " | ...... | 67,200 | 83,500 | ...... | ...... | 19.88 |
| 1142* | 3-66 | " | 57,500 | 67,200 | 91,900 | 7,136 | 11.00 | 26.88 |
| Av'rage. | ...... | ...... | 57,500 | 66,933 | 89,067 | ...... | ...... | 24.83 |
| 1147 | 1-66 | $1\frac{5}{16}$ | 56,200 | 67,500 | 87,900 | 3,616 | 5.85 | 23.21 |
| 1148 | 2-66 | " | 60,000 | 67,500 | 96,200 | 4,710 | 7.35 | 29.78 |
| 1149* | 3-66 | " | 56,200 | 68,500 | 83,300 | 2,913 | 4.90 | 17.81 |
| Av'rage. | ...... | ...... | 57,467 | 67,833 | 89,133 | 4,163 | 6.60 | 23.60 |
| 1154 | 1-66 | 1 | 58,709 | 67,800 | 100,900 | 5,164 | 8.05 | 32.77 |
| 1155* | 2-66 | " | 63,700 | 68,500 | 101,900 | 4,909 | 7.45 | 32.77 |
| 1156* | 3-66 | " | 58,700 | 68,200 | 101,400 | 3,360 | 5.30 | 32.77 |
| Av'rage. | ...... | ...... | 60,367 | 68,167 | 101,400 | 5,164 | 8.05 | 32.77 |
| 1161 | 1-66 | $\frac{5}{8}$ | 63,800 | 73,800 | 106,800 | 3,566 | 4.85 | 30.90 |
| 1162 | 2-66 | " | 67,100 | 72,200 | 92,000 | 2,657 | 3.75 | 19.81 |
| 1163 | 3-66 | " | 60,600 | 75,500 | 104,800 | 3,670 | 5.00 | 27.96 |
| Av'rage. | ...... | ...... | 63,833 | 73,833 | 101,200 | 3,298 | 4.53 | 26.22 |

All the numbers marked thus * broke outside of the initial marks.

## B.—UNTREATED IRON.

| Lab. No. | Original Mark. | Diameter. Inches | Elastic Limit. | Modulus of Rupture per square inch of | | Modulus of Resilience. | Percentage of | |
|---|---|---|---|---|---|---|---|---|
| | | | | Original Section. | Fractured Section. | | Extension. | Redct'n of area at Fracture. |
| 1136* | 4-66 | 2 7/16 | 29,800 | 46,900 | 67,800 | 8,777 | 20.55 | 31.55 |
| 1137 | 5-66 | " | 26,200 | 46,900 | 74,700 | 11,081 | 25.55 | 37.85 |
| 1138* | 6-66 | " | 29,800 | 46,400 | 68,500 | 7,710 | 18.20 | 82.09 |
| Av'age | ......... | ......... | 28,600 | 46,733 | 70,333 | 11,081 | 26.25 | 33.83 |
| 1143 | 4-66 | 2 1/16 | 28,200 | 48,500 | 69,000 | 8,232 | 19.25 | 29.63 |
| 1144 | 5-66 | " | 28,200 | 48,900 | 81,300 | 12,717 | 28.75 | 39.80 |
| 1145 | 6-66 | " | ............ | 48,100 | ............ | ............ | ............ | ............ |
| Av'age | ......... | ......... | 28,200 | 48,500 | 75,150 | 10,475 | 24.00 | 34.72 |
| 1150* | 4-66 | 1 3/8 | 24,300 | 50,300 | 83,000 | 9,272 | 22.00 | 39.68 |
| 1151* | 5-66 | " | 24,300 | 50,300 | 80,000 | 6,497 | 15.65 | 37.16 |
| 1152* | 6-66 | " | 24,300 | 50,300 | 81,500 | 10,141 | 22.60 | 38.31 |
| Av'age | ......... | ......... | 24,300 | 50,300 | 81,500 | ............ | ............ | 38.37 |
| 1157* | 4-66 | 1 1/16 | 26,100 | 47,300 | 81,500 | 7,946 | 21.75 | 41.91 |
| 1158* | 5-66 | " | 28,900 | 47,300 | 77,600 | 9,882 | 23.60 | 38.97 |
| 1159* | 6-66 | " | 28,100 | 47,600 | 79,900 | 9,718 | 25.30 | 40.45 |
| Av'age | ......... | ......... | 27,700 | 47,400 | 79.667 | ............ | ............ | 40.44 |
| 1164* | 4-66 | 3/4 | 29,200 | 50,100 | 72,200 | 8,001 | 16.55 | 30.61 |
| 1165 | 5-66 | " | 29,200 | 43,600 | 69,900 | 8,761 | 19.35 | 37.66 |
| 1166* | 6-66 | " | 29,200 | 50,800 | 78,500 | 10,032 | 20.95 | 37.66 |
| Av'age | ......... | ......... | 29,200 | 48,167 | 73,533 | 8,761 | 19.35 | 35.31 |

*All numbers marked thus * broke outside of the initial marks.

## C.—COLD-ROLLED AND ANNEALED IRON.

| Lab. No. | Original Mark. | Diameter. Inches | Elastic Limit. | Modulus of Rupture per square inch of | | Modulus of Resilience. | Percentage of | |
|---|---|---|---|---|---|---|---|---|
| | | | | Original Section. | Fractured Section. | | Extension. | Reduc'n of area at Fracture. |
| 1139* | X | 2 7/16 | 31,400 | 46,300 | 75,400 | 6,076 | 14.25 | 38.80 |
| 1146* | XII | 2 | 31,800 | 49,600 | 78,400 | 5,619 | 12.50 | 36.78 |
| 1153 | ......... | 1 5/16 | 31,600 | 49,500 | 86,900 | 4,927 | 9.50 | 43.09 |
| 1160 | ......... | 1 | 32,700 | 50,900 | 75,600 | 5,857 | 12.65 | 32.77 |
| 1167* | ......... | 5/8 | 33,600 | 48,700 | 76,100 | 6,777 | 15.80 | 36.00 |

*All those test-pieces whose numbers are marked with an asterisk (*), broke outside of the scale, and therefore the extensions given in the table are not the ultimate extensions, and the Moduli of Resilience are only those due to the extensions measured, and are not comparable.

# LOT No. 2.

All specimens of this Lot were turned to their respective diameters from bars 2 inches in diameter.

## D—COLD-ROLLED IRON.

| Lab. No. | Original Mark. | Diameter. Inches | Elastic Limit. | Modulus of Rupture per square inch of | | Modulus of Resilience. | Percentage of | |
|---|---|---|---|---|---|---|---|---|
| | | | | Original Section. | Fractur'd Section. | | Extension. | Reduction of area at frac're. |
| 1168* | ......... | 1¾ | 63,900 | 66,900 | 94,800 | 3,877 | 6.00 | 29.44 |
| 1169 | ......... | 1½ | 56,600 | 68,500 | 95,500 | 4,930 | 7.65 | 28.30 |
| 1170 | ......... | 1 | 56,700 | 60,600 | 86,200 | 3,794 | 6.55 | 31.12 |

## E—UNTREATED IRON.

| 1171 | ......... | 1¾ | 30,900 | 48,700 | 83,100 | 14,120 | 30.00 | 41.38 |
| 1172 | ......... | 1½ | 33,500 | 49,500 | 82,700 | 11,567 | 25.70 | 40.18 |
| 1173* | ......... | 1 | 26,000 | 47,900 | 78,700 | 8,997 | 21.30 | 39.14 |

*All test-pieces whose numbers are marked with an asterisk (*), broke outside of the scale, therefore the extensions given in the table opposite the respective numbers are not the ultimate extensions, and the Moduli of Resilience are only those due to the measured extensions, and therefore neither are comparable, but serve to show that the total extensions and ultimate Moduli of Resilience for those particular specimens do not fall below the values given in the table.

## LOT No. 3.

All specimens of this Lot were turned to their respective diameters from bars 2 inches in diameter.

### F—COLD-ROLLED IRON.

| Lab. No. | Diam. In | Elastic Limit. | Modulus of Elasticity. | Modulus of Rupture per sq. in. of Original Section. | Fractu'd Section. | Modulus of Resilience. Elastic. | Ultimate. | Percentage of Extension. | Reduction of area of frac. |
|---|---|---|---|---|---|---|---|---|---|
| 1105 A | ⅞ | 54,900 | 26,781,000 | 65,850 | 96,000 | 109.76 | 7200 | 11.07 | 31.85 |
| 1105 B | ¾ | 56,600 | 27,829,000 | 65,640 | 93,000 | 69.59 | 5892.5 | 9.00 | 29.66 |
| 1105 C | ⅝ | 55,400 | 25,743,000 | 66,650 | 90,600 | 71.47 | 5640 | 9.22 | 26.47 |
| 1105 D | ½ | 56,000 | 28,989,000 | 66,200 | 91,600 | 76.47 | 5072.5 | 8.14 | 27.76 |
| 1105 E | ⅜ | 54,300 | 22,261,000 | 63,400 | 99,000 | 163.13 | 4247.5 | 7.29 | 28.89 |
| 1105 F | ¼ | 50,900 | 35,553,000 | 64,660 | 91,800 | 51.16 | 1945 | 3.43 | 29.60 |

### G—UNTREATED IRON.

| Lab. No. | Diam. In | Elastic Limit. | Modulus of Elasticity. | Modulus of Rupture per sq. in. of Original Section. | Fractu'd Section. | Modulus of Resilience. Elastic. | Ultimate. | Percentage of Extension. | Reduction of area of frac. |
|---|---|---|---|---|---|---|---|---|---|
| 1104 A | ⅞ | 23,300 | 23,860,000 | 58,450* | 122,400* | 14.91 | 12167.5* | 26.30 | 34.53 |
| 1104 B | ¾ | 23,800 | 25,679,000 | 49,380 | 79,400 | 18.77 | 8607.5 | 21.57 | 37.85 |
| 1104 C | ⅝ | 24,100 | 30,863,000 | 50,520 | 90,100 | 18.09 | 1087.5 | 24.57 | 43.94 |
| 1104 D | ½ | 23,900 | 30,150,000 | 50,980 | 93,600 | 26.33 | 7757.5 | 18.57 | 40.42 |
| 1104 E | ⅜ | 20,800 | 27,038,000 | 52,540 | 97,700 | 26.25 | 8500 | 20.57 | 46.24 |
| 1104 F | ¼ | 22,400 | 33,317,000 | 42,980 | 78,400 | 23.52 | 653.5 | 16.93 | 47.28 |

*The Moduli of Rupture of this test-piece, and in a less degree the Modulus of Resilience, have been unduly augmented by giving the test-piece an opportunity to recover, after being strained greatly beyond its Elastic Limit. The Modulus of Rupture would very probably not have much exceeded 50,000 pounds, and the Modulus of Resilience would undoubtedly have fallen below 12,000 foot-pounds.

## II—SUMMARY OF RESULTS.

### TRANSVERSE TESTS OF COLD-ROLLED SHAFTING.

*Results reduced to Standard ; bar 1 inch sq. and 22 inches long.*

| Lab. No. | Diam. in inches | Elasticity. | | Resilience in foot-pounds. | | Modulus of Maximum Resistance. | Maximum Load. | Deflection for Maximum Load. |
|---|---|---|---|---|---|---|---|---|
| | | Limit. lbs. | Modulus. | Elastic. | When D=2 inches. | | | |
| 1109A′ | $2\frac{7}{16}$ | 3,165 | 29,998,000 | 42.19 | 575.12 | 133,480 | 4,040 | 2.30 |
| 1110A′ | " | 3,165 | 28,195,000 | 42.19 | 573.49 | 134,640 | 4,080 | 2.27 |
| 1111A′ | " | 3,165 | 30,843,000 | 40.67 | 679.91 | 134,640 | 4,030 | 1.99 |
| Av'age. | ........ | 3,165 | 29,678,700 | 41.68 | 576.17 | 134,250 | 4,050 | 2.19 |
| 1109B′ | 2 | 2,950 | 27,896,000 | 45.47 | 560.54 | 136,000 | 4,200 | ...... |
| 1110B′ | " | 3,050 | 26,452,000 | 54.00 | 545.12 | 135,200 | 4,120 | ...... |
| 1111B′ | " | 3,050 | 28,256,000 | 52.87 | 550.75 | 135,000 | 4,100 | ...... |
| Av'age. | ........ | 3,017 | 27,535,000 | 50.78 | 552.14 | 135,400 | 4,140 | ...... |
| 1109C′ | $1\frac{5}{16}$ | 2,450 | 26,625,000 | 26.45 | 494.72 | 114,890 | 3,482 | 2.20 |
| 1110C′ | " | 2,600 | 25,655,000 | 32.97 | 500.34 | 117,620 | 3,569 | 2.56 |
| 1111C′ | " | 2,550 | 27,671,000 | 34.76 | 503.46 | 118,390 | 3,584 | 2.56 |
| Av'age. | ........ | 2,530 | 26,650,300 | 31.39 | 499.50 | 116,970 | 3,545 | 2.44 |
| 1109D′ | 1 | 3,000 | 27,094,000 | 43.74 | 571.15 | 136,740 | 4,160 | 2.13 |
| 1110D′ | " | 3,056 | 27,163,000 | 44.47 | 582.19 | 136,690 | 4,143 | 2 |
| 1111D′ | " | 3,056 | 27,094,000 | 47.22 | 576.46 | 133,430 | 4,076 | 1.71 |
| Av'age. | ........ | 3,037 | 27,117,000 | 45.14 | 576.60 | 135,620 | 4,126 | 1.95 |
| 1109E′ | $\frac{9}{8}$ | 2,850 | 27,443,000 | 55.80 | 545.62 | 133,090 | 4,033 | 1.53 |
| 1110E′ | " | 2,850 | 27,094,000 | 56.99 | 540.21 | 133,090 | 4,033 | 2.00 |
| 1111E′ | " | 2,850 | 27,094,000 | 58.18 | 529.27 | 133,090 | 4,033 | 2.00 |
| Av'age. | ........ | 2,850 | 27,210,300 | 56 99 | 538.40 | 133,090 | 4,033 | 1 84 |

D=Deflection in inches.

# I—SUMMARY OF RESULTS.

## TRANSVERSE TESTS OF HOT-ROLLED SHAFTING.

*Results reduced to Standard; bar 1 inch sq., 22 inches long.*

| Lab. No. | Diameter of bar. Inches | Elasticity. | | Resilience in foot-pounds. | | Modulus of Maximum Resistance. | Maximum Load. | Deflect'n under maxim'm load. Inches. |
|---|---|---|---|---|---|---|---|---|
| | | Limit. Po'nds | Modulus. | Elastic | When D=2″ | | | |
| 1106A | 2.54 | 1,500 | 25,422,000 | 10.81 | 339.74 | 85,470 | 2,590 | 2.66 |
| 1107A | " | 1,450 | 25,659,000 | 7.55 | 338.90 | 90,600 | 2,745 | 2.87 |
| 1108A | " | 1,500 | 26,146,000 | 8.44 | 342.66 | 88,890 | 2,690 | 2.80 |
| Av'rage. | ......... | 1,487 | 25,742,300 | 8.77 | 340.43 | 88,320 | 2,675 | 2.78 |
| 1106B | 2.08 | 1,375 | 28,197,000 | 9.74 | 282.49 | 71,590 | 2,170 | 3.00 |
| 1107B | " | 1,350 | 27,592,000 | 8.44 | 286.00 | 74,080 | 2,245 | 4.00 |
| 1108B | " | 1,350 | 28,404,000 | 8.59 | 282.49 | 73,210 | 2,220 | 3.5 |
| Av'rage. | ......... | 1,358 | 28,064,000 | 8.92 | 283.66 | 72,960 | 2,212 | 3.5 |
| 1106C | 1.36 | 1,300 | 28,143,000 | 7.52 | 265.56 | 65,900 | 1,975 | 2.58 |
| 1107C | " | 1,330 | 28,624,000 | 8.64 | 270.16 | 64,990 | 1,959 | 1.91 |
| 1108C | " | 1,350 | 28,624,000 | 9.56 | 269.54 | 64,540 | 1,934 | 2.19 |
| Av'rage. | ......... | 1,327 | 28,130,300 | 8.57 | 268.42 | 65,143 | 1,956 | 2.23 |
| 1106D | 1.04 | 1,720 | 25,782,000 | 14.37 | 317.92 | 80,690 | 2,455 | 4 |
| 1107D | " | 1,720 | 27,233,000 | 14.58 | 322.09 | 82,180 | 2,490 | 4 |
| 1108D | " | 1,750 | 25,782,000 | 14.52 | 326.56 | 84,670 | 2,565 | 4 |
| Av'rage. | ......... | 1,740 | 26,265,700 | 14.49 | 322.19 | 82,513 | 2,503 | 4 |
| 1106E | 0.665 | 1,750 | 28,528,000 | 21.88 | 298.13 | 72,390 | 2,194 | 3.25 |
| 1107E | " | 1,750 | 28,705,000 | 21.88 | 292.71 | 72,390 | 2,194 | 3.75 |
| 1108E | " | 1,750 | 28,343,000 | 20.41 | 294.59 | 72,390 | 2,194 | 3.5 |
| Av'rage. | ......... | 1,750 | 28,525,300 | 21.39 | 295.14 | 72,390 | 2,194 | 3.5 |

D=Deflection in inches.

# K—SUMMARY OF RESULTS.

## TRANSVERSE TESTS OF COLD-ROLLED AND ANNEALED SHAFTING.

*Results reduced to Standard; bar 1 inch sq., 22 inches long.*

| Lab. No. | Diam. of Bar. Inches. | Elasticity. | | Resilience in foot-pounds. | | Modulus of Maximum Resistance. | Maximum Load. | Deflection under maximum load. |
|---|---|---|---|---|---|---|---|---|
| | | Limit. | Modulus. | Elastic. | When D=2″. | | | |
| 1128A' | $2\frac{7}{16}$ | 1,800 | 30,539,000 | 11.25 | 394.31 | 94,790 | 2,870 | 2.56 |
| 1128B' | 2 | 1,850 | 27,463,000 | 15.41 | 354.74 | 84,670 | 2,570 | 2.52 |
| 1128C' | $1\frac{5}{16}$ | 1,900 | 27,188,000 | 19.00 | 352.90 | 82,620 | 2,503 | 2.93 |
| 1128D' | 1 | 1,700 | 25,845,000 | 15.58 | 319.17 | 77,030 | 2,334 | 3 |
| 1128E' | $\frac{5}{8}$ | 1,808 | 26,687,000 | 24.75 | 333.96 | 78,480 | 2,378 | 4 |

# TABLE I.—WORKING AND BREAKING LOADS

## OF

## COLD-ROLLED AND OF HOT-ROLLED IRON SHAFTING.

*Deduced from Tests made at the Mechanical Laboratory, Department of Engineering, Stevens Institute of Technology.*

| MATERIAL. | Diameter. Inches | Shafts one foot long, between bearings; loaded at the middle. Loads—pounds. Maximum. | Proof. | Working. | Deflection per 100 lbs. of load. Inches. | Shafts fixed at one end, loaded at the other. Lever-arm, one foot. Loads in pounds. Maximum. | Proof. | Working. | Deflection per 100 lbs of load. Inches. |
|---|---|---|---|---|---|---|---|---|---|
| Cold-Rolled | 2 7/16 | 65,000 | 49,000 | 24,500 | 0.00010 | 15,750 | 12,250 | 6,100 | .00072 |
| Hot-Rolled | 2 9/16 | 47,000 | 26,000 | 13,000 | .00012 | 11,750 | 6,500 | 3,250 | .00095 |
| Cold-Rolled | 2 | 37,000 | 28,000 | 13,000 | .00018 | 9,250 | 6,500 | 3,250 | .00144 |
| Hot-Rolled | 2 1/16 | 22,000 | 13,000 | 6,500 | .00016 | 5,500 | 3,250 | 1,600 | .00128 |
| Cold-Rolled | 1 5/8 | 8,600 | 6,300 | 3,150 | .00093 | 2,150 | 1,570 | 780 | .00744 |
| Hot-Rolled | 1 8 | 5,400 | 8,800 | 1,900 | .00077 | 1,350 | 950 | 470 | .00616 |
| Cold-Rolled | 1 | 4,400 | 3,200 | 1,600 | .0025 | 1,100 | 800 | 400 | .02242 |
| Hot-Rolled | 1 1/16 | 3,000 | 2,700 | 1,050 | .00243 | 750 | 520 | 260 | .01944 |
| Cold-Rolled | 5/8 | 1,060 | 760 | 380 | .01782 | 260 | 190 | 90 | .14256 |
| Hot-Rolled | 3/4 | 700 | 500 | 250 | .01447 | 170 | 120 | 60 | .11562 |

NOTE.—By "Maximum Load" is meant the greatest load which the shaft will sustain before becoming completely distorted or broken. By "Proof Load" is meant the maximum load which the shaft can sustain without taking an appreciable permanent set. The "Working Load" is that to which the shaft may be repeatedly subjected without injury. If subjected to shock, however, only one-half of the figures in the table is allowable.

The table gives the load for a length of one foot; to obtain the load for any other length we have $W' = \frac{W}{L}$; where $W'$ is the required load, W, that given in the table, and L is the length of the proposed shaft in feet. The deflections in the table are for each 100 pounds on a shaft one foot long; for any other load less than the proof, and for any other length, we have $D' = \frac{D P L^3}{100}$; where $D'$ is the desired deflection, P, any load in pounds less than the Proof, L the length of the shaft for which the deflection is sought, and D the deflection given in the table.

# ABSTRACT OF A REPORT

ON

# TESTS OF COLD-ROLLED IRON

MADE IN THE

## AUTOGRAPHIC RECORDING TESTING MACHINE.

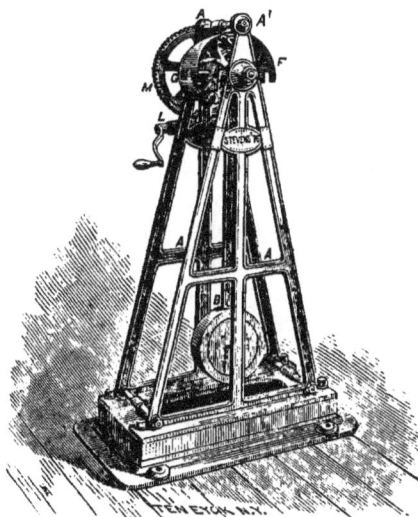

TESTING MACHINE FOR TORSION.

The machine employed in testing by torsional stress is known as the Autographic Recording Testing Machine. It is shown in the accompanying engraving, as built and used in the Mechanical Laboratory of the Stevens Institute of Technology.

It consists of two strong cast iron wrenches, facing each other, with a space of $1\frac{1}{4}$ inches between their jaws. They rotate on independent journals, placed in the same line in the frames, *AA;* the latter are bolted to a heavy bed-plate, which gives it the required stability. One of the wrenches is provided with an arm, 4.5 feet in length, at the lower end of which is attached a heavy weight, *B;* the other wrench has keyed to it a worm-wheel, *M,* engaging with the worm, *L,* which is set in motion by means of a crank. In this manner a very slow and quite uniform motion can be obtained.

Both wrenches are provided with lathe-centres directly opposite each other and in the common axis of rotation. The specimen to be tested is placed upon the lathe-centres, which hold it in line while it is being secured in the jaws of the wrenches by means of steel wedges inserted from opposite sides.

On the shaft of the wrench carrying the worm-wheel there is fastened a brass drum, *G,* which rotates with it, while to the other wrench is fastened a pencil-holder which allows the point of the pencil to move on the surface of the drum and is guided by the stationary curve, *F,* of brass, in such a manner that its position on the drum indicates the number of foot-pounds of moment exerted by the arm and weight, at any instant.

Supposing a test-piece to be placed in the machine, the operator turns the crank, *L,* with a uniform velocity which gives a slow and a very steady motion to the wrench connected with the worm-wheel, which is transmitted through the test-piece to the wrench carrying the weighted arm. The latter is moved by the force transmitted through the test-piece through an arc which is a measure of the resistance to torsion offered by the test-piece, and is recorded simultaneously with the angle of torsion by the pencil upon a diagram-sheet fastened upon the drum for the purpose.

The drum is of such a diameter that the circumference is 36 inches, which, when divided into tenths, make 360 divisions, each of which is representative of one degree. The guide-curve is a curve of sines, which insures the position of the pencil on the drum always such that it marks an ordinate proportional to the moment of the arm and weight at every instant during the test.

In the machine employed in the Mechanical Laboratory of the Stevens Institute of Technology, each inch of ordinate denotes 100 foot-pounds of moment to have been transmitted through the test-piece, and each inch of abscissa indicates 10 degrees of torsion.

The friction of the machine is not recorded by the machine, but is added in calculating the results given in the tables.

By means of this machine the metal tested is compelled to tell its own story and to give a permanent and graphical representation of its strength, elasticity, and every other quality which is brought into play during its test, and to exhibit the characteristic peculiarities.

## DEFINITIONS AND EXPLANATIONS.

Before passing to the discussion of the results of the torsion tests themselves, it is necessary, in order to avoid misunderstanding, to define some of the more important terms, and to explain the methods by which the results given in Table M and to which data reference is frequently made in the discussions, were derived.

1. *The Modulus of Torsional Elasticity* is the ratio of the distorting force to the amount of angular distortion which it produces, in a test-piece of which the length, the polar Moment of Inertia and the lever-arm of the applied force are each unity; thus

$$G = \frac{P\,A\,L\dagger}{\theta\,I_p} \quad \dots \dots \dots \dots \dots \dots \dots \dots \dots (3)$$

Where

G = Modulus of Torsional Elasticity,
P = the applied force,
L = length of test-piece,
A = length of arm of P,
$\theta$ = angle of distortion produced by P,
$I_p$ = the polar Moment of Inertia,

$$= \frac{\pi\,r^4}{2}, \ r \text{ being the radius of the test-piece.}$$

And $\pi = 3.1416$.

It must be remembered that in calculating the Moduli of Elasticity, P should always be taken well within the Elastic Limit.

† Wood's Resistance of Materials, new edition, p. 206.

2. *Resilience.* The actual Torsional Resiliences, both Elastic and Ultimate, are given in the tables also. The Torsional Resilience is calculated by means of the formula,

$$W = \frac{A M}{R} + \frac{L F}{R}$$

which is derived as follows:

Let W=the required Resilience, in foot-pounds,

      A=area in square inches included in the diagram up to the point of rupture,

      L=length of the base line in inches,

      M=the number of foot-pounds of moment represented by each inch of ordinate, supposing no friction, and

      R=radius of the drum in inches, we have

$$Y = \frac{A}{L} = \text{mean ordinate of diagram,}$$

P=Y M=value of mean ordinate, in foot-pounds, or pounds applied at distance of one foot from the axis of the jaws of the machine, which is the mean force exerted to rupture the test-piece ;

$$S = \frac{L}{12} = \text{distance moved through, in feet, by pencil, and since}$$

the pencil point is at a distance equal to the radius of the drum from the axis, the distance through which the mean force acts is

$(\frac{R}{12}$=radius of drum in feet): one foot :: $[\frac{L}{12} = \substack{\text{distance through} \\ \text{which mean force acts.}}] : \frac{L}{R}$

Therefore, still supposing no friction, the work done will be the product of the mean force into the distance through which it acts, and,

$$\frac{P L}{R} = \frac{A}{L} M \times \frac{L}{R} = \frac{A M}{R}$$

The friction F, of the journal, previously referred to, assists this mean force, and is taken as acting at a distance of one foot from the axis ; it therefore acts through the same distance $\frac{L}{R}$, and other work is therefore expended equal to $\frac{F L}{R}$

so that the total work performed is

$$W = \frac{A}{R}\frac{M}{R} \times \frac{F}{R}\frac{L}{R}$$

The values of the Elastic Resilience, or the work expended in straining the material to its Elastic Limit, are determined in the same manner as those for the Ultimate Resilience. To find the Resilience for the Elastic Limit, the area $a$, of the initial portion of the diagram up to the Limit of Elasticity is measured, as also the abscissa $l$, corresponding to that point which when substituted in the formula above given, M, F and R remaining the same, we have the following:

Resilience within the Elastic Limit $= a\frac{M}{R} + l\frac{F}{R}$

3. *Moduli of Torsional Resistance.* * The Moduli of Resistance, Proof and Maximum, as given in Table R, are represented in foot-pounds of Moment, *i. e.*, in pounds of stress acting upon a lever-arm one foot long.

*The Proof Modulus,* $A(=\frac{M}{D^3})$, represents the number of foot-pounds of stress required to strain a round bar, whose diameter $(D)$ is unity,† up to its Elastic Limit.

*The Maximum Modulus,* $A'(=\frac{M'}{D^3})$, is the greatest resistance, in foot-pounds, offered by the same bar while being ruptured. The ultimate resistance, or the resistance at the point of breaking, is not always a maximum.

*The Modulus of Elastic Stiffness,* $A''(=\frac{ML}{D^3\theta^z})$, is the moment in foot-pounds necessary to twist a specimen of the given material, whose length $(L)$ and diameter $(D)$ are both unity, through an angle of one degree.

---

*The above Moduli of Torsional Resistance are calculated according to Rankine; Machinery and Mill Work, p. 501.

† The inch is here taken as the unit of measure.

4. *Ductility by Torsion.* The relative ductility by torsion is given in the table under the heads "$\theta$, Final," and "Ratio of Extension, Ultimate." The former is the number of degrees of torsion, and the latter the extension of an external fibre at the point of rupture.

The extension is that of an external fibre of the metal, originally lying parallel with the axis, and usually exceeds in value that obtained by tension. It is calculated for the standard test-piece, whose length is one inch, and diameter 0.625 inch, by means of the formula:

$$S = \sqrt{1 + A^2 \times 0.00002974775} - 1$$

in which S is the extension and A the total angle of torsion.

—

## DISCUSSION OF RESULTS OF TESTS BY TORSION.

All the test-pieces were cut from round bars $1\frac{1}{8}$ in. diameter and dressed to the standard size in the workshop of the Mechanical Laboratory of the Stevens Institute of Technology. Twelve specimens are here reported upon; 4 of the untreated iron, 6 of cold-rolled iron and 2 were cold-rolled and annealed iron. In the bar from which Nos. 1547B and 1548B were cut, the process of cold-rolling was carried farther than in the others.

### UNTREATED IRON.

The average Modulus of Resistance at the Elastic Limit is 381.8 foot-pounds, and the Maximum is 932.93 foot-pounds of Moment. The average Elastic Resilience is 0.898, and the Ultimate Resilience is 733.13 foot-pounds of work. Excepting specimen No. 1219, the Moduli of Elasticity are over 20,000,000, which is very high for torsion. The Modulus of Elastic Stiffness is also very high, the average being 1151.49 foot-pounds of Moment. The extension of the external fibre varies between 30 and 80 per cent. The strain-diagrams Nos. 1218, 1219, 1546A and 1547A, on Plate XVI, show graphically the behavior of these specimens under torsional stress.

The high elastic stiffness ; the smoothness with which the curves turn on passing the elastic limit ; their regularity of rise ; their altitude and their high torsional angles, show the metal to be an excellent material even as it leaves the common mill. It is stiff and strong, yet ductile, and is both uniform in quality and homogeneous in structure.

No. 1218, for example, gives one of the most perfect strain-diagrams that is to be found among the hundreds preserved in the portfolios of the Mechanical Laboratory of the Stevens Institute of Technology. It may be taken as a typical diagram for the finest quality of merchant bar-iron.

Nos. 1219, 1546A and 1547A are seen to be a trifle more fibrous, as is shown by the slight depression of the line after passing the Elastic Limit E, and somewhat less ductile, but they are also very excellent specimens of hot-rolled iron.

### COLD-ROLLED IRON.

The high Elastic Limit and Ultimate Stiffness, which is one of the most prominent features of cold-rolled iron, is as well marked in torsion as in tension and in transverse resistance. The Proof Moment is nearly double that of the untreated iron. The average Proof Modulus of Nos. 1203C, 1207C, 1547A and 1547B is 699.30 foot-pounds; that of Nos. 1548A and 1548B is higher still, it being 922.15 foot-pounds of Moment. The average Modulus of Ultimate Strength of the former is 1035.55, and of the latter it is 1094.67 foot-pounds of Moment. The difference in Ultimate Strength is not so great as in the Elastic Limit. The average ductility of the external fibre is 20.17 per cent. Nos. 1203C and 1548B are exceptionally low in ductility ; the former showed a flaw very plainly near the fracture, which reduced its strength and elastic stiffness.

The average Elastic Resilience is 3.49 foot-pounds of work, nearly four times that of the untreated iron. The Ultimate Resilience is 561.12 foot-pounds of work.

The Moduli of Elasticity and Elastic Stiffness are not quite as high as in the untreated iron.

### Cold-Rolled and Annealed Iron.

Annealing the cold-rolled iron reduces it very nearly to the state of the untreated iron in its properties to resist torsional as well as other stresses, but leaves the material more homogeneous as to strain. This is shown by the smoothness of the curves, Nos. 1804 A and 1808A, which do not exhibit the counterflexure generally observed in strain-diagrams produced by specimens of untreated iron. No. 1804A retains the characteristics conferred upon it by cold-rolling, sufficiently to show plainly its origin.

### The Peculiar Action of Cold-Rolling.—Earlier Experiments.

The writer had, as early as the year 1873, obtained, for his own satisfaction, a set of test-pieces of cold-rolled iron which were tested in the Autographic Testing Machine. The strain-diagrams were described, and a *fac-simile* printed in a paper read before the American Society of Civil Engineers, in April, 1874, as follows:

" No. 85 is a singular illustration of the effects of what is probably a peculiar modification of internal strain. It seems to have no characteristics in common with any other metal examined. Its diagram would seem to show a perfect homogeneousness as to strain, and a remarkable deficiency of homogeneity in structure. It begins to exhibit the indications of an Elastic Limit at $a$, under a torsional movement of 110 foot-pounds, or an apparent tensile stress of 33,000 pounds per square inch, and then rises at once by a beautifully regular curve, to very nearly its maximum at 18°, and 176 foot-pounds. The maximum is finally reached at 130°, and thence the line slowly falls until fracture takes place at 195°. The maximum resistance seems to be very exactly 60,000 pounds to the square inch. Its maximum elongation for exterior fibres is about 0.23. The Resilience taken at the Elastic Limit is far higher than with common iron, and it is seen that this metal, in many respects, may compete with steel. Its elasticity is seen to remain constant wherever taken.

" This singular specimen was a piece of ' cold-rolled' iron. It is probably really far from homogeneous as to strain, but its artificially produced strains are symmetrically distributed about its axis, and being rendered perfectly uniform throughout each of the concentric cylinders into which it may be conceived to be divided, the effect, so far as this test, or so far as its application as shafting, for example, is concerned, is that of perfect homogeneousness.

" The apparently great deficiency of homogeneousness in structure is readily explained by an examination of the pieces after fracture ; they are fibrous, and have a grain as threadlike as oak ; their condition is precisely what is shown by the diagram, and the metal itself is as anomalous as its curve."

It was by these strain-diagrams, and at the time here referred to, that the real character of the peculiar and important change produced by cold-rolling was first discovered, and shown to be a remarkable elevation of the original elastic limit of the material.

The elaborate series of investigations described in the report of which this is a partial abstract, have fully confirmed the deductions then made by the writer, and it is here shown for the first time that the exaltation of the primitive elastic limit by strain may be carried to such an extent as to make it equal to eighty per cent. and more, of the breaking load, a proportion which is not obtainable by any other process known to the writer. This fact has as much interest and importance when viewed from the scientific as from the practical side.

In the strain-diagrams of Nos. 1548 A and 1548 B, the modification produced in the cold-rolling mill is most strikingly shown. The stiffness indicated by the steepness of the initial line rising to E, the height of the elastic limit—which is seen to approximate closely to the maximum resistance of the piece—and the increased strength of the iron, are apparent to the least observing eye. The ductility is seen to be somewhat reduced, but still remains available for all ordinary applications, and is greater than that of some hot-rolled iron which finds a ready market and has a high reputation.

These diagrams not only exhibit the facts just given with perfect distinctness, but also show that the final rolling cannot be done at a high temperature if the improvement sought is to be made thoroughly satisfactory and permanent. The study and measurement of these diagrams will probably prove more instructive and satisfactory to the majority of readers than the examination of the tabulated figures.

# GENERAL CONCLUSIONS.

From a study of the accompanying strain-diagrams and the appended tables, as well as from what has previously been stated, the following general conclusions, already drawn, are fully corroborated.

(1.) The process of cold-rolling produces a very marked change in the physical properties of the iron thus treated.

(a.) It increases the tenacity from 25 to 40 per cent., and the resistance to transverse stress from 50 to 80 per cent.

(b.) It elevates the Elastic Limit under torsional as well as tensile and transverse stresses, from 80 to 125 per cent.

(c.) The Modulus of Elastic Resilience is elevated from 300 to 400 per cent. The Elastic Resilience to transverse stress is augmented from 150 to 425 per cent.

(2.) Cold-rolling also improves the metal in other respects :

(a.) It gives the iron a smooth, bright surface, absolutely free from the scale of black oxyde unavoidably left when hot-rolled.

(b.) It is made exactly to gauge and for many purposes requires no further preparation.

(c.) In working the metal, the wear and tear of the tools are less than with hot-rolled iron, thus saving labor and expense in fitting.

(d.) The cold-rolled iron resists stresses much more uniformly than does the untreated metal. Irregularities of resistance exhibited by the latter do not appear in the former ; this is more particularly true for transverse stress, as is shown by the smoothness of the strain-diagrams produced by the cold-rolled bars.

(e.) This treatment of iron produces a very important improvement in uniformity of structure, the cold-rolled iron excelling common iron in its uniformity in density from surface to centre, as well as in its uniformity of strength from outside to the middle of the bar.

(3.) This great increase of strength, stiffness, Elasticity and Resilience is obtained at the expense of some ductility, which diminishes as the tenacity increases. The Modulus of Ultimate Resilience of the cold-rolled iron is, however, above 50 per cent. of that of the untreated iron.

Cold-rolled iron thus greatly excels common iron in all cases where the metal is to sustain maximum loads without permanent set or distortion.

*Comparing the Autographic Strain-diagrams,* we see evidence:

(1.) That the curves exhibit the same peculiarities that were there also observed when testing these metals by transverse stress, and by tension. The diagrams of the cold-rolled iron, after the Elastic Limit is passed, gradually falls into a horizontal line; while those of the untreated metal turn abruptly and generally show a counterflexure in the curve, just beyond the Elastic Limit.

(2.) That the diagrams of the annealed cold-rolled iron still retain some of the characteristics of those of the unannealed.

(3.) That the result of the treatment of the metal is the elevation of the Elastic Limit more or less nearly to the limit of strength observed at final rupture and the change of the method of passing the Elastic Limit, making that change far less abrupt, and giving a smoother and more symmetrical curve than that noted on the strain-diagrams of the hot-rolled metal.

Very Respectfully,

R. H. THURSTON.

# TABLE M.

### — OF —

## SUMMARY OF RESULTS OF TESTS BY TORSION

### UNTREATED, COLD-ROLLED, AND COLD-ROLLED AND ANNEALED IRON.

*Standard Test-piece 1 inch long between shoulders, and 0.625 inch in diameter.*

| MATERIAL | Lab. No. | Stresses in Foot-Pounds of Moment. | | Angle of Torsion. | | Ratio of Extension of an External Fibre. | | Modulus of Resistance in Foot-Pounds of Moment. | | Modulus of Elasticity. | Elastic Stiffness. | | Actual Resilience in Foot Pounds of Work. | |
| --- | --- | --- | --- | --- | --- | --- | --- | --- | --- | --- | --- | --- | --- | --- |
| | | At Elastic Limit. $M$ | Maximum. $M'$ | At Elastic Limit. $\theta_e$ | Final. $\theta_m$ | Proof. | Ultimate. | Proof. $A=\dfrac{M}{D^3}$ | Maximum. $A'=\dfrac{M'}{D^3}$ | $G$ | Actual $\dfrac{M}{\theta_e}$ | Modulus $A_r=\dfrac{M\,L}{D^5\,\theta_m}$ | Elastic. $W$ | Ultimate $W'$ |
| **Untreated Iron.** | 1218 | 90.36 | 234.22 | 0.60 | 270.20 | .000009 | .780962 | 370.12 | 9?9.37 | 23,193,000 | 160.60 | 1579.1 | 0.49 | 985.37 |
| | 1219 | 86.31 | 228.14 | 1.10 | 201.40 | .000019 | .485473 | 353.53 | 934.46 | 9,359,000 | 78.46 | 822.70 | 0.86 | 685.54 |
| | 1546A | 100.25 | 240.25 | 1.30 | 215.90 | .000028 | .544871 | 410.60 | 984.00 | 20,067,000 | 77.12 | 808.45 | 1.49 | 774.39 |
| | 1546B | 93.25 | 208.25 | 0.70 | 157.70 | .0000105 | .319018 | 382.97 | 853.90 | 20,891,000 | 133.20 | 1396.70 | 0.75 | 487.20 |
| **Cold-Rolled Iron.** | 1203C | 156.21 | 225.10 | 2.15 | 73.80 | .00071 | .077972 | 639.84 | 922.01 | 7,514,000 | 72.66 | 761.80 | 2.49 | 271.91 |
| | 1207O | 156.21 | 271.70 | 2.00 | 197.50 | .000066 | .469818 | 639.84 | 1112.88 | 12,855,000 | 78.10 | 818.90 | 2.80 | 860.87 |
| | 1647A | 189.25 | 260.25 | 2.10 | 120.00 | .000067 | .195143 | 775.10 | 1065.90 | 20,265,000 | 90.12 | 949.57 | 4.17 | 517.60 |
| | 1547B | 181.25 | 264.25 | 2.10 | 140.40 | .000067 | .259522 | 742.40 | 1041.40 | 19,317,000 | 86.30 | 904.87 | 4.52 | 594.10 |
| | 1548A | 225.25 | 267.25 | 3.50 | 97.80 | .000186 | .133374 | 922.60 | 1094.67 | 16,493,000 | 64.30 | 674.71 | 7.76 | 435.70 |
| | 1548B | 226.25 | 267.25 | 3.70 | 71.90 | .000207 | .074145 | 921.70 | 1094.67 | 14,954,000 | 61.20 | 641.08 | 8.12 | 317.00 |
| **Cold-Rolled and Annealed Iron.** | 1204A | 86.31 | 235.23 | 1.90 | 238.90 | .000055 | .642498 | 363.52 | 963.56 | 5,335,000 | 45.42 | 476.20 | 1.20 | 856.64 |
| | 1208A | 92.39 | 224.05 | 1.20 | 204.80 | .000024 | .499238 | 378.43 | 917.87 | 9,198,000 | 76.99 | 807.30 | 1.01 | 672.53 |

# ABSTRACT OF A REPORT

ON

# COLD-ROLLED FINGER-BARS,

## FOR MOWING MACHINES.

In the preceding pages I have reported in full upon a very extensive and complete series of tests of *Cold-Rolled Shafting* and the iron from which it is made.

The conclusions there arrived at in reference to the beneficial effects of the process of cold-rolling, as practiced at the American Iron Works of Jones & Laughlins, Pittsburgh, Pa., are general, and would seem to apply equally well to other forms of iron similarly treated.

To determine whether this is true with regard to finger bars, as made at the American Iron Works, and to compare the cold-rolled iron finger-bars with those of steel made in the ordinary way, was the object of a series of tests, of which the following is a brief abstract :

Tables N and O are summaries of results obtained by tension and transverse stress respectively, by studying which the great superiority of the cold-rolled iron, in strength and ultimate stiffness, is seen at a glance. The process being carried further, the difference is even more marked than in the cold-rolled shafting.

The Modulus of Rupture is nearly double, and the Elastic Limit is more than three times that of the untreated iron ; the Modulus of Elastic Resilience is nearly tenfold. Ductility is reduced, but the Ultimate Resilience does not suffer in the same ratio. In tension it is a full third of that of the untreated iron, and under transverse stress it falls but very slightly below the latter in Ultimate Resilience. This metal is therefore peculiarly well adapted for finger-bars, which, generally, are only called upon to overcome transverse resistance.

In Table P is given a summary of the results of tests of eight finger-bars by transverse stress. The untreated bars were ¾ inch thick ; to make them directly comparable the results were reduced to those for a bar of the same dimensions as the cold-rolled bars. Comparing the reduced figures with those of the tests of cold-rolled bars the great superiority of the latter is at once evident. Even in direct comparison with the untreated finger-bars, which are 50 per cent. heavier than the cold-rolled, the latter surpass the former. The Elastic Resilience is nearly three times as great, and the Ultimate Resilience is also considerably higher. The untreated bars reach the Elastic Limit with a deflection of 1 inch, whereas the cold-rolled bars can be bent 2.75 inches before any appreciable set occurs.

Manufacturers of reaping and mowing machines are well aware of the fact that for finger-bars a material of great strength and high elasticity is required. Cold-rolled iron is, therefore, much better adapted for the purpose than is untreated iron.

## COMPARISON BETWEEN COLD-ROLLED IRON AND STEEL FINGER-BARS.

The great Strength, Elasticity and Ultimate Stiffness of cold-rolled iron, as shown by the numerous tests already made, and particularly the great regularity with which it yields under stress, justifies the conclusion that for many purposes it may be used, with advantage, in place of steel. It is inferred that in consequence of the peculiar character of its resisting qualities, cold-rolled iron is especially adapted for the most important members of harvesting machines.

To determine the exact relative value of cold-rolled iron and steel finger-bars, by direct comparison, a somewhat extended series of tests were made—a summary of the results of which is found in Table Q.

From the table it is seen that the cold-rolled iron bars are from fifteen to twenty-five per cent. stronger than those of steel. The Resilience is also much higher. To break the cold-rolled bars it was necessary to bend them back and forth, 14 inches each way many times; one of the bars was bent 10 inches each way 23 times before it ruptured. The permanent sets for large deflections is much less with the cold-rolled iron than with the steel bars. (See Table Q under the head of Permanent Sets.)

In addition to the above, the cold-rolled iron finger-bars exhibit a regularity of product and a smoothness of finish, which is greatly in their favor. The work of fitting is also much more easily performed upon the cold-rolled iron than upon the steel bars.

TABLES R AND S are here appended as giving for comparison the results of a series of independent supplementary tests of cold-rolled and untreated irons which may be useful for further comparison.

IN CONCLUSION, it may be said that cold-rolled iron finger-bars are well calculated to take the place of those made of steel, rolled in the ordinary manner, and that many other parts of the working mechanism, as well as the framework of harvesting machines, might, with advantage, be replaced by cold-rolled iron, thereby securing what is very desirable in such machines—a combination of maximum strength with minimum weight.

Very Respectfully,

H. R. THURSTON.

# TABLE N.

## SUMMARY OF RESULTS OF TESTS BY TENSION

— O F —

### COLD-ROLLED, AND COLD-ROLLED AND ANNEALED FINGER-BAR IRON.

| MATERIAL. | Lab. No. | Elastic Limit. | Modulus of Elasticity. | Moduli of Rupture per Square inch of | | Moduli of Resilience. | | Extension in Per Cent. | Reduction of Area In Per Cent. |
|---|---|---|---|---|---|---|---|---|---|
| | | | | Original Section. | Fractured Section. | Elastic. | Ultimate. | | |
| UNTREATED IRON. | 1484A | 26,000 | 26,669,000 | 46,400 | 68,400 | 22.10 | 9,365.00 | 22.83 | 32.10 |
| | 1485A | 26,000 | 26,669,000 | 48,080 | 70,000 | 23.40 | 10,320.00 | 24.00 | 31.32 |
| | 1484B | 23,500 | 23,226,000 | 46,800 | 55,400 | 21.15 | 8,287.50 | 20.00 | 15.62 |
| | 1485B | 23,500 | 23,226,000 | 47,200 | 53,000 | 19.98 | 8,812.50 | 20.83 | 10.95 |
| | 1484C | 23,300 | 26,570,000 | 45,525 | 54,600 | 18.64 | 6,447.50 | 16.18 | 20.48 |
| | 1485C | 22,000 | 24,341,000 | 45,100 | 55,080 | 20.70 | 7,012.50 | 17.33 | 20.00 |
| | 1484D | 23,500 | 25,000,000 | 46,280 | 61,900 | 17.62 | 9,260.00 | 22.50 | 28.41 |
| | 1485D | 23,500 | 25,925,000 | 46,280 | 65,700 | 16.45 | 10,570.00 | 25.67 | 29.73 |
| COLD-ROLLED IRON. | 1486A | 84,000 | 24,350,000 | 86,690 | 99,800 | 210.00 | 3,972.50 | 4.83 | 13.17 |
| | 1487A | 83,000 | 24,659,000 | 85,710 | 101,500 | 202.00 | 5,475.00 | 6.67 | 15.62 |
| | 1486B | 79,000 | 22,652,000 | 82,790 | 87,100 | 197.60 | 2,085.00 | 2.83 | 4.93 |
| | 1487B | 76,800 | 20,042,000 | 83,510 | 86,000 | 192.00 | 1,420.00 | 1.83 | 1.70 |
| COLD-ROLLED AND ANNEALED IRON. | 1486C | 31,000 | 22,769,000 | 43,550 | 53,600 | 29.08 | 4,182.00 | 11.00 | 18.83 |
| | 1487C | 31,000 | 22,652,000 | 47,080 | 56,500 | 29.08 | 6,672.00 | 15.88 | 16.72 |

# TABLE O.

## SUMMARY OF RESULTS OF TESTS BY TRANSVERSE STRESS

OF

### UNTREATED, COLD-ROLLED, AND COLD-ROLLED AND ANNEALED FINGER-BAR IRON.

RESULTS ARE ALL REDUCED TO THOSE FOR A STANDARD BAR 1 INCH SQUARE AND 22 INCHES BETWEEN SUPPORTS.

| Dimensions. | Material | Laboratory Number. | Moduli of Elasticity. | Maximum Resistance. | Resilience in Foot Pounds. | | Loads in Pounds. | | Deflections in inches at | |
|---|---|---|---|---|---|---|---|---|---|---|
| | | | | | Elastic. | Ultimate. | Elastic Limit. | Maximum | Elastic Limit. | Max. Load. |
| Rectangular section. — Dist. bet. Supports, 22 in. — 1 x 0.75 inches. | UNTREATED IRON. | 1121A* | 24,269,000 | 65,570 | 9.43 | 894.79 | 1,300 | 1,987 | 0.35 | 3.82 |
| | | 1121B* | 21,802,000 | 61,990 | 7.08 | 839.17 | 1,125 | 1,878 | 0.32 | 3.81 |
| | | 1121C | 22,652,000 | 55,440 | 6.67 | 764.38 | 1,050 | 1,680 | 0.30 | 3.10 |
| | | 1121D | 22,082,000 | 57,850 | 6.83 | 775.42 | 1,100 | 1,753 | 0.30 | 3.07 |
| 1 x 0.5 inches. | COLD-ROLLED IRON. | 1120A* | 24,716,000 | 137,280 | 45.62 | 594.58 | 3,000 | 4,160 | 0.72 | 1.26 |
| | | 1120B | 25,297,000 | 132,000 | 34.87 | 822.16 | 2,700 | 4,000 | 0.62 | 1.25 |
| | COLD-ROLLED AND ANNEALED IRON. | 1120C | 21,819,000 | 58,740 | 10.50 | 810.00 | 1,400 | 1,780 | 0.36 | 3.31 |

*All the numbers marked thus (*) were tested edgewise, i. e., the load was applied in a direction perpendicular to that of the pressure of the rolls of the mill.

# TABLE P.

## SUMMARY OF RESULTS OF TESTS BY TRANSVERSE STRESS

### OF

## COLD-ROLLED AND UNTREATED IRON FINGER-BARS.

| Material. | Laboratory Number. | Moduli of | | Resilience in Foot Pounds. | | Loads in Pounds. | | Deflections in Inches at | |
|---|---|---|---|---|---|---|---|---|---|
| | | Elasticity. | Maximum Resistance. | Elastic. | Ultimate. | Elastic Limit. | Maximum. | Elastic Limit. | Max. Load. |
| UNTREATED IRON. | 1118A | 27,265,000 | 49,700 | 44.12 | 1101.67 | 1055 | 1210 | 1.00 | 6.85 |
| | 1118B | 25,428,000 | 46,500 | 40.43 | 1039.17 | 990 | 1140 | 1.00 | 4.88 |
| | 1118C | 26,656,000 | 47,100 | 37.95 | 1052.83 | 990 | 1160 | 0.92 | 7.85 |
| | 1118D | 25,582,000 | 45,800 | 34.63 | 979.83 | 880 | 1100 | 0.94 | 5.87 |
| | 1118A′ | 27,265,000 | 49,700 | 27.71 | 495.83 | 476 | 552 | 1.40 | 8.33 |
| | 1118B′ | 25,428,000 | 46,500 | 24.19 | 470.83 | 430 | 520 | 1.35 | 5.91 |
| | 1118C′ | 26,656,000 | 47,100 | 23.83 | 470.35 | 440 | 530 | 1.30 | 9.38 |
| | 1118D′ | 25,582,000 | 45,800 | 20.00 | 445.17 | 380 | 502 | 1.21 | 7.28 |
| COLD-ROLLED IRON. | 1119A | 26,096,000 | 115,700 | 106.31 | 1055.67 | 950 | 1300 | 2.70 | 6.88 |
| | 1119B | 25,684,000 | 113,800 | 110.00 | 1045.33 | 960 | 1280 | 2.75 | 6.85 |
| | 1119C | 26,866,000 | 113,200 | 93.37 | 997.67 | 900 | 1270 | 2.50 | 5.81 |
| | 1119D | 26,585,000 | 108,800 | 86.00 | 977.42 | 860 | 1220 | 2.40 | 5.85 |

Dimensions.*

Thickness, 0.75 in.  Thickness, 0.5 in.  Breadth at the Middle, 3.15 inches.  Distance between Supports, 48 inches.

* The breadth was not precisely the same for all the bars.

# TABLE Q.

## SUMMARY OF RESULTS OF TESTS BY TRANSVERSE STRESS OF COLD-ROLLED IRON AND STEEL FINGER-BARS.

| Dimensions | Material | Lab. No. | Moduli of | | Resilience in Foot Lbs. | | Loads in Pounds. | | Deflections at | | Per'nt Sets in in's after deflecting | | |
|---|---|---|---|---|---|---|---|---|---|---|---|---|---|
| | | | Elasticity. | Max. Resis. | Elastic. | Ultimate. | Elast. Lim | Maximum | Elast. Lim | Max. Load. | 3.0" | 3.5" | 1.0" |
| Breadth at the Middle 3.45 inches, Thickness, 0.5 inches. | Cold Rolled Iron. | 1227 | 28,222,000 | 116,400 | 126.04 | 1,151.67 | 1100 | 1400 | 2.80 | 5.34 | 0.30 | 0.60 | 0.85 |
| | | 1228 | 27,009,000 | 112,700 | 133.00 | 1,152.67 | 1140 | 1400 | 2.75 | 5.86 | 0.30 | 0.58 | 0.85 |
| | | 1229 | 28,005,000 | 115,600 | 121.50 | 1,139.33 | 1080 | 1390 | 2.70 | 5.85 | 0.33 | 0.00 | 0.85 |
| | Steel. | 1230 | 29,685,000 | 84,900 | 97.75 | 209.00 | 1020 | 1050 | 2.30 | 2.54 | 0.70 | 1.10 | 1.70 |
| | | 1231 | 30,211,000 | 98,800 | 127.83 | 1,038.50 | 1180 | 1220 | 2.60 | 4.37 | 0.50 | 1.00 | 1.50 |
| | | 1232 | 29,517,000 | 90,100 | 104.77 | 957.67 | 1070 | 1110 | 2.35 | 2.91 | 0.72 | 1.25 | 1.70 |
| Breadth at the Middle, 3.7 inches. Thickness, 0.375 inches | Cold Rolled Iron. | 1233 | 27,233,600 | 115,300 | 103.12 | 640.83 | 660 | 820 | 3.75 | 6.98 | 0.08 | 0.14 | 0.30 |
| | | 1234 | 26,997,000 | 113,900 | 109.08 | 633.83 | 680 | 810 | 3.85 | 6.49 | 0.04 | 0.15 | 0.25 |
| | | 1235 | 27,285,000 | 115,300 | 109.08 | 643.67 | 680 | 820 | 3.86 | 5.96 | 0.09 | 0.18 | 0.31 |
| | Steel. | 1236 | 28,616,000 | 81,400 | 74.82 | 506.67 | 600 | 625 | 3.00 | 3.99 | 0.20 | 0.30 | 1.00 |
| | | 1237 | 29,780,000 | 95,100 | 93.52 | 567.00 | 670 | 700 | 3.35 | 4.01 | 0.05 | 0.25 | 0.60 |
| | | 1238 | 28,682,000 | 86,300 | 79.44 | 512.33 | 618 | 640 | 3.10 | 4.01 | 0.10 | 0.51 | 0.80 |
| Breadth at the Middle 2.87 inches. Thickness, 0.14 inch. | Steel. | 1239 | 28,378,000 | 92,900 | 90.42 | 583.50 | 705 | 720 | 3.20 | 3.51 | 0.20 | 0.60 | 1.00 |
| | | 1240 | 28,749,000 | 92,900 | 86.74 | 579.00 | 690 | 720 | 3.00 | 4.03 | 0.20 | 0.60 | 1.09 |
| | | 1241 | 28,259,000 | 93,100 | 93.20 | 589.33 | 695 | 730 | 3.00 | 4.03 | 0.27 | 0.60 | 1.22 |
| | Cold Rolled Iron. | 1242 | 27,419,000 | 112,700 | 88.00 | 637.67 | 660 | 825 | 3.20 | 6.03 | 0.20 | 0.31 | 1.20 |
| | | 1243 | 26,983,000 | 116,100 | 96.16 | 716.83 | 725 | 910 | 3.15 | 5.54 | 0.18 | 0.25 | 0.48 |
| | | 1244 | 27,017,000 | 115,200 | 99.69 | 709.33 | 725 | 880 | 3.30 | 5.53 | 0.16 | 0.30 | 0.55 |

Length between Supports, 48 inches.

# TABLE R.

## SUMMARY OF RESULTS OF TESTS BY TRANSVERSE STRESS,

*Of Jones & Laughlins' Common and Cold-Rolled Irons.*

| MATERIAL. | Lab. No. | Elastic Limit. | Modulus of Elasticity. | Modulus of Rupture. | Resilience. | | Resistance. | | | | Deflect for Maximum Load. | REMARKS. |
|---|---|---|---|---|---|---|---|---|---|---|---|---|
| | | | | | Elastic. | When D*=4″ | When D*=4″ | When D*=6″ | When D*=9″ | Maximum. | | |
| Cold-Rolled Iron. | 1190 | 2800 | 25,418,000 | 135,000 | 42·8 | 1067·27 | 3350 | (a) | (a) | 3430 | 3″ | (a) Loads not observed. |
| | 1191 | 2600 | 25,970,000 | 132,000 | 37·7 | 1081·23 | 3280 | 1900 | 1230 | 3360 | 3 | |
| Average, | | 2700 | 25,691,500 | 133,600 | 40·25 | 1049·25 | 3315 | | | 3395 | 3 | |
| Common Iron, | 1216 | 1800 | 26,620,000 | 73,200 | 9·75 | 550·61 | 1860 | 1770 | 1380 | 1860 | 4 | |
| | 1217 | 1850 | 27,729,000 | 74,000 | 11·75 | 552·90 | 1880 | 1760 | 1400 | 1880 | 4 | |
| Average, | | 1325 | 27,174,500 | 73,600 | 10·75 | 551·75 | 1870 | 1765 | 1390 | 1870 | 4 | |

*D represents total deflection in inches.

†The figures given in this table are not reduced to the standard bar, but are derived from the test-pieces which were 1¾ inches in diameter and 22 inches between supports.

# TABLE S.

## SUMMARY OF RESULTS OF TESTS BY TENSILE STRESS,

### Of Jones & Laughlins' Irons, Common, Cold-Rolled, and Cold-Rolled and Annealed.

| Material. | Lab. No. | Modulus of Elasticity. | Elastic Limit. | Modulus of Rupture. Original Section. | Fractured Section. | Resilience. Elastic. | Ultimate. | Percentage of Exten- sion. | Reduc'n of Area. | Loads, Total. Maxi- mum. | Final. | Remarks. |
|---|---|---|---|---|---|---|---|---|---|---|---|---|
| Cold-Rolled | 1196 | 26,374,000 | 60,000 | 69,000 | 105,600 | 93.00 | 6,677 | 10 | 34.70 | 34,600 | 28,600 | |
| | 1199 | 26,087,000 | 58,000 | 69,000 | 106,600 | 89.00 | 7,120 | 10.83 | 34.70 | 34,500 | ........ | Final load not observed. |
| Average. | | 26,230,500 | 59,000 | 69,000 | 105,600 | 91.00 | 6,898.5 | 10.42 | 34.70 | 34,500 | ........ | |
| Cold-Rolled and Annealed. | 1195 | 25,918,000 | 30,000 | 52,000 | 95,100 | 31.96 | 12,487 | 26.66 | 45.32 | 26,000 | 22,950 | |
| | 1200 | 24,691,000 | 34,000 | 51,200 | 90,600 | 23.80 | 11,074 | 23.33 | 43.46 | 25,000 | 23,650 | |
| Average. | | 25,304,500 | 33,000 | 51,600 | 92,850 | 27.88 | 11,780.5 | 25 | 44.39 | 25,500 | 23,300 | R somewhat high, because the test-piece was allowed to rest 15 min. under load of 25,000 pounds. |
| Common Iron. | 1221 | 24,641,000 | 28,000 | 53,600 | 93,200 | 22.80 | 10,617 | 23.33 | 42.5 | 26,800 | 23,700 | |
| | 1222 | 25,918,000 | 32,000 | 52,000 | 90,200 | 18.88 | 11,440 | 25.16 | 42.32 | 26,000 | ........ | Final load not observed. |
| | 1225 | 26,667,000 | 28,000 | 52,600 | 90,000 | 21.75 | 11,020 | 24.17 | 41.55 | 26,300 | 24,000 | |
| | 1226 | 23,857,000 | 32,000 | 51,800 | 86,100 | 20.80 | 11,756 | 25.67 | ........ | 25,900 | 23,500 | |
| Average. | | 25,270,750 | 30,000 | 52,500 | 89,870 | 21.06 | 11.183 | 24.68 | ........ | 26,250 | 23,733 | |

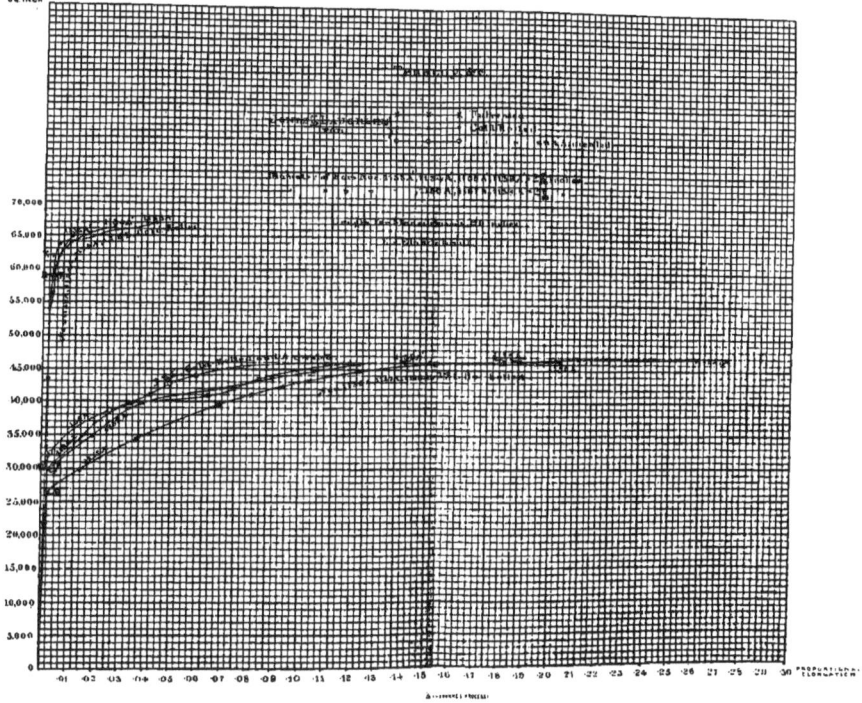

PLATE Nº 1.

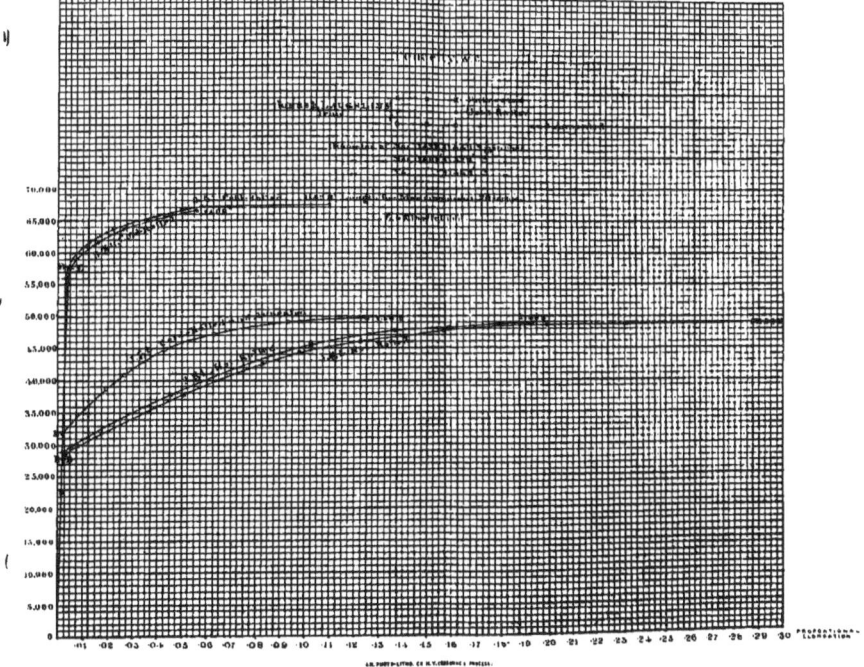

o
:

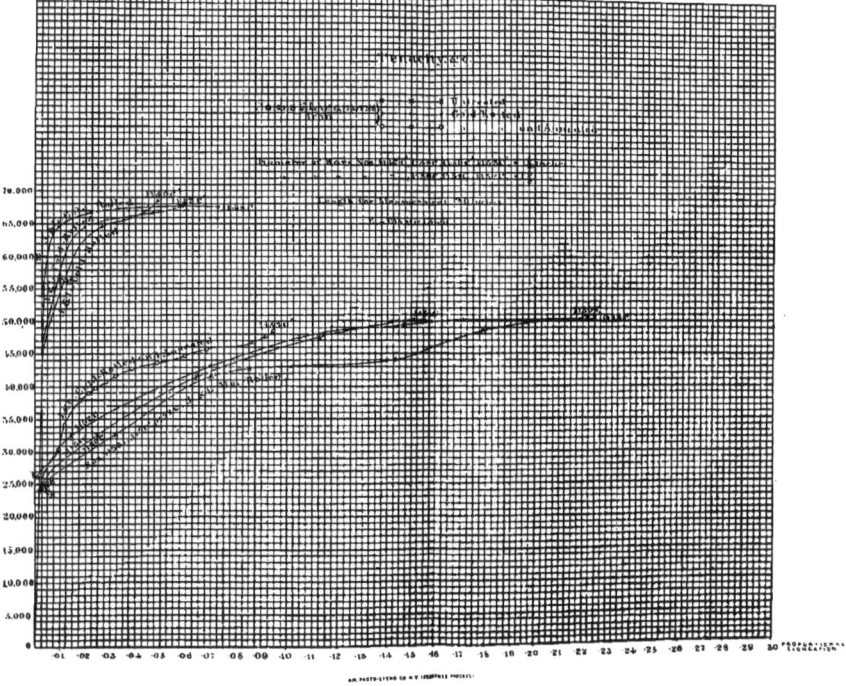

PLATE Nº III

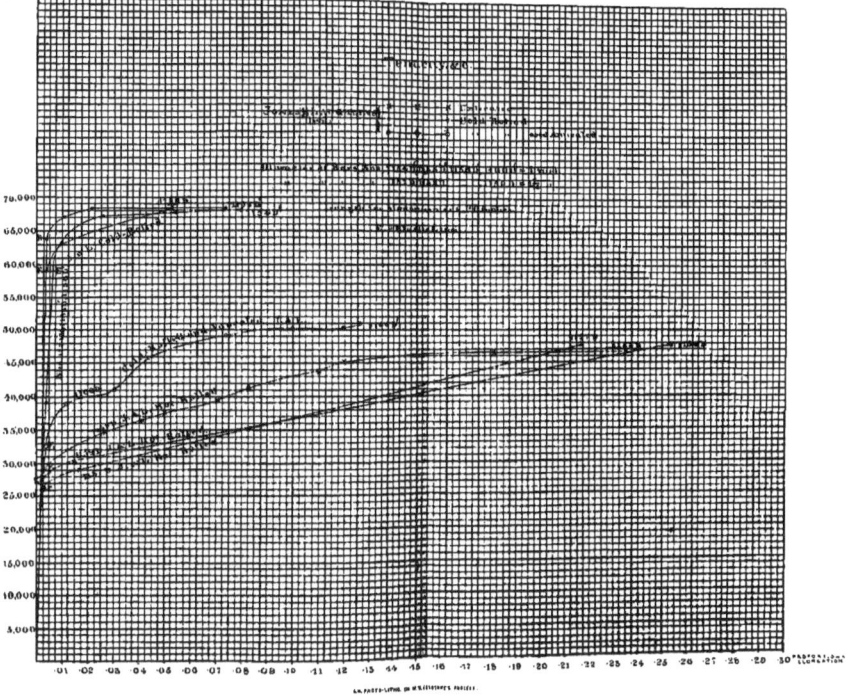

PLATE Nº IV.

(

6

5(

50.

45

40.

35,0

50,0

25,0

20,00

15,00

10.000

5,000

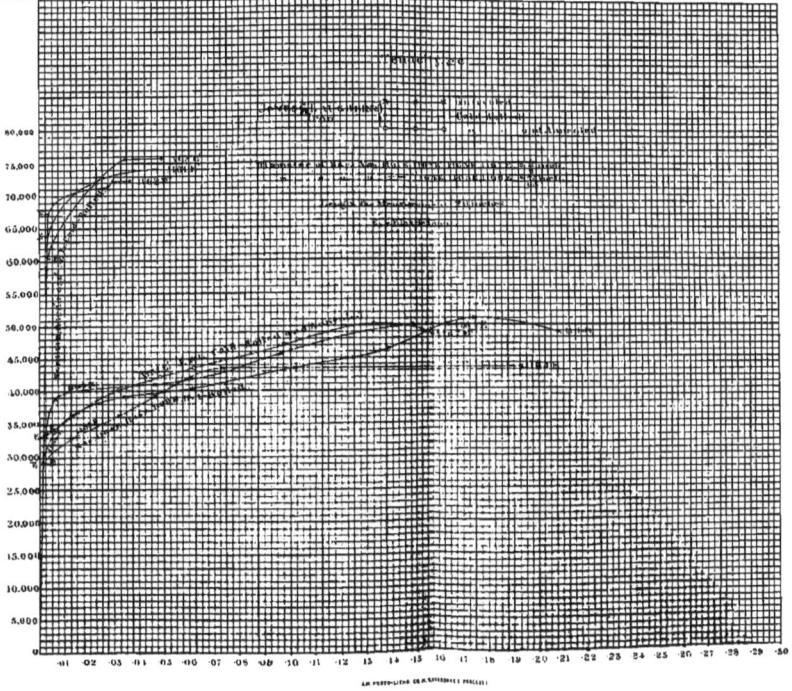

PLATE Nº V

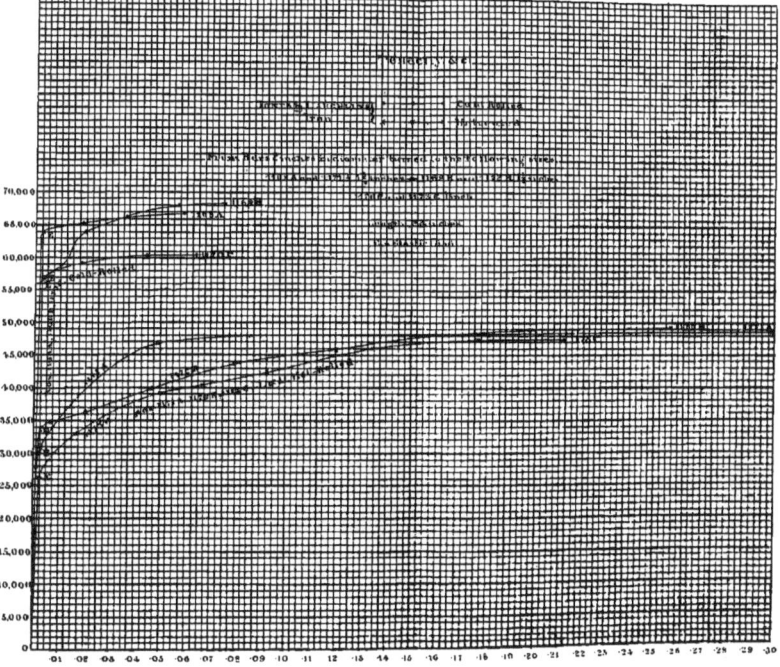

PLATE No. VI.

PLATE Nº IX.

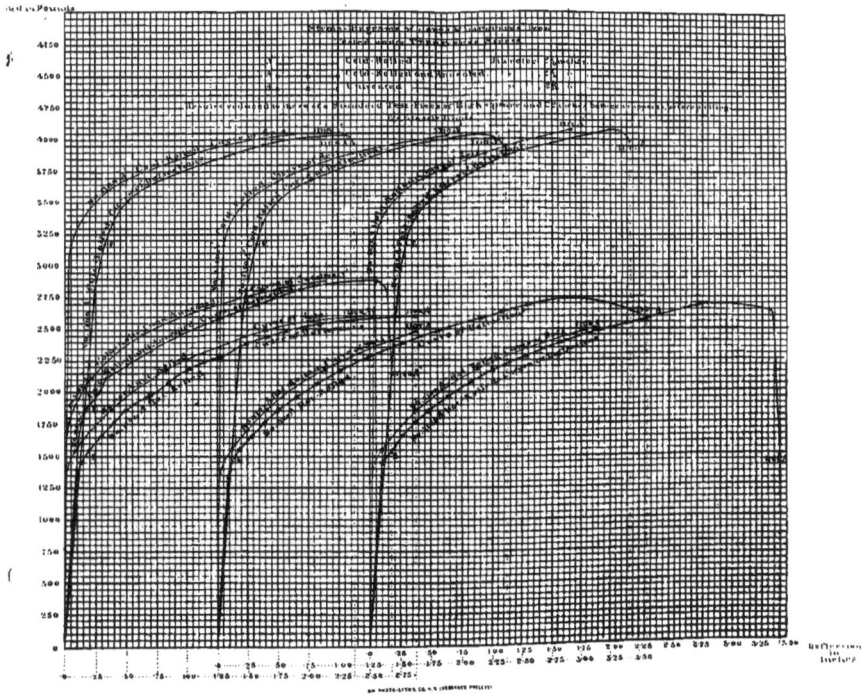

PLATE Nº X.

ection
bi
ches.

Load in Pounds.

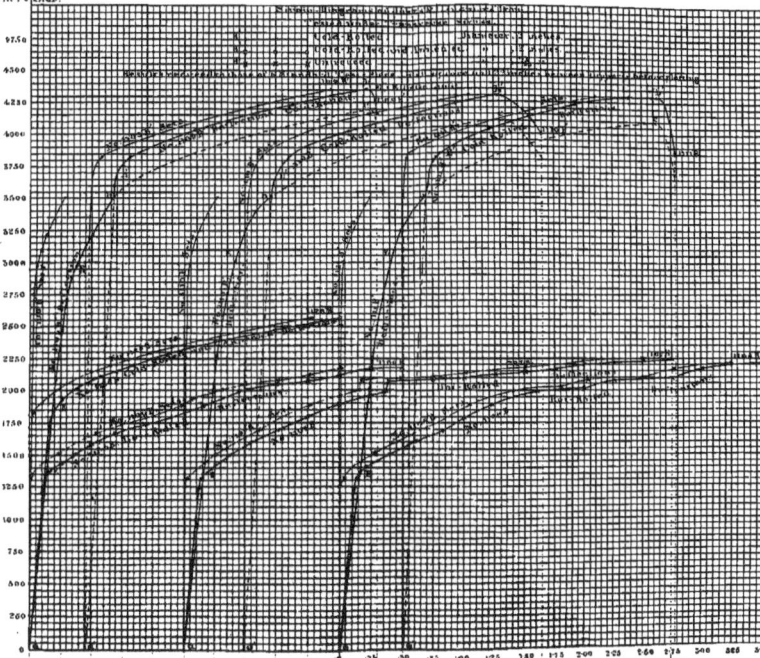

Deflection in Inches

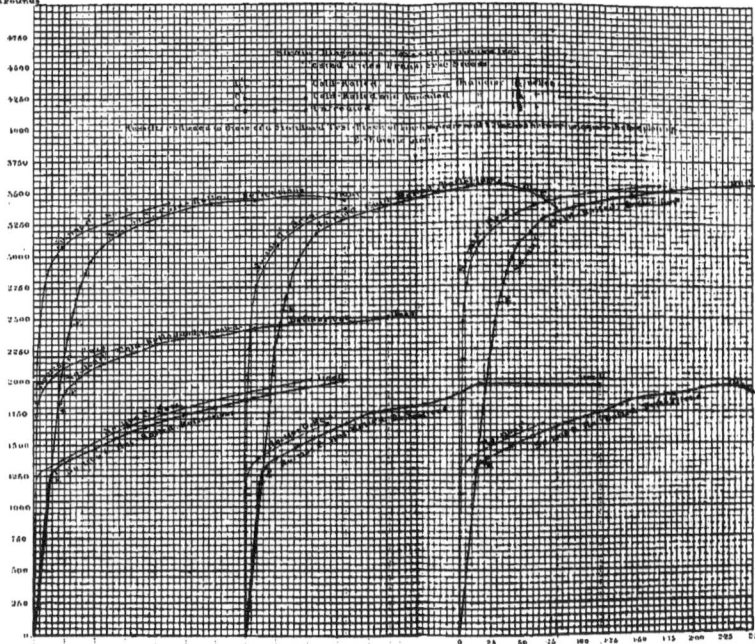

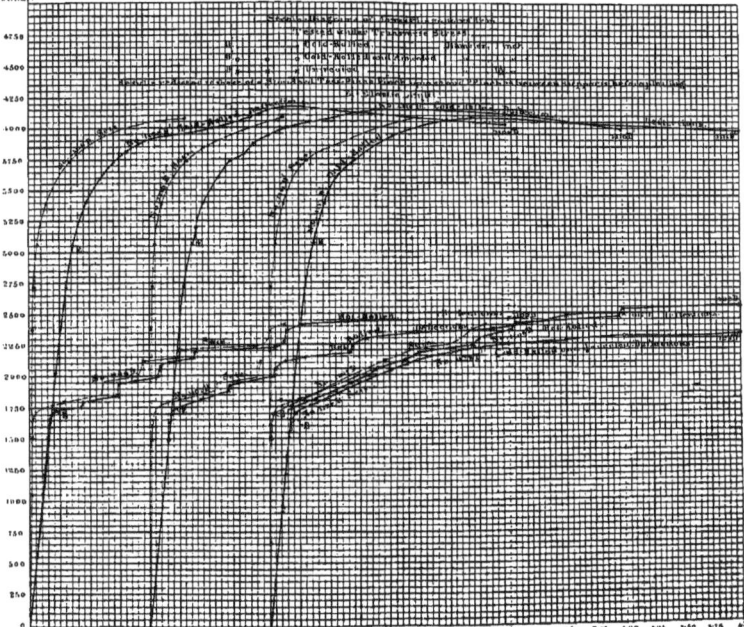

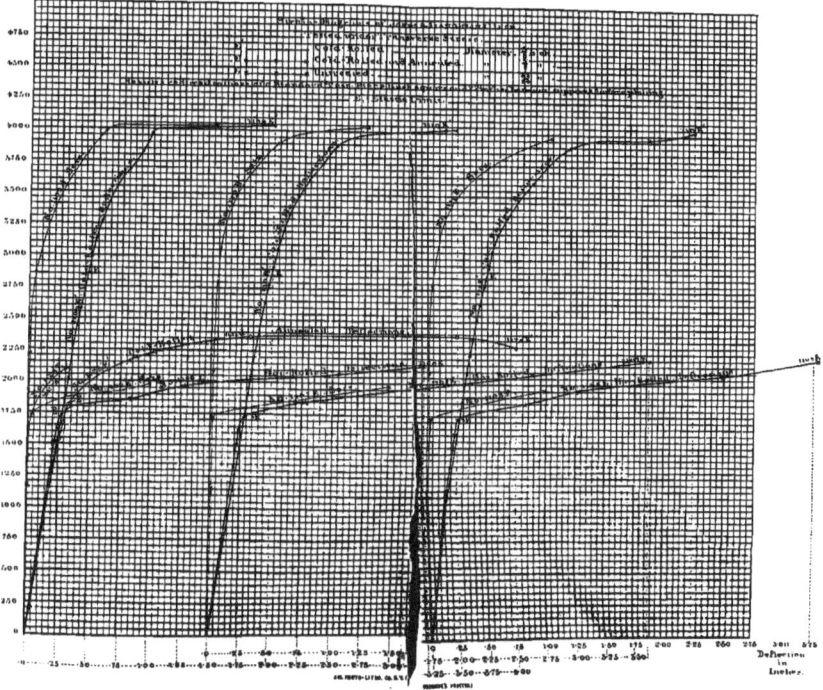
PLATE Nº XIV.

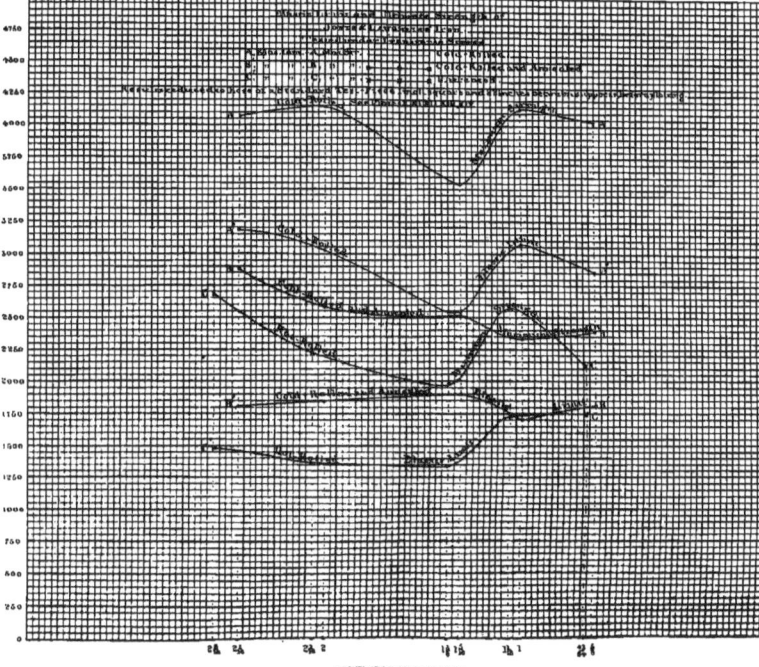

PLATE Nº XV.

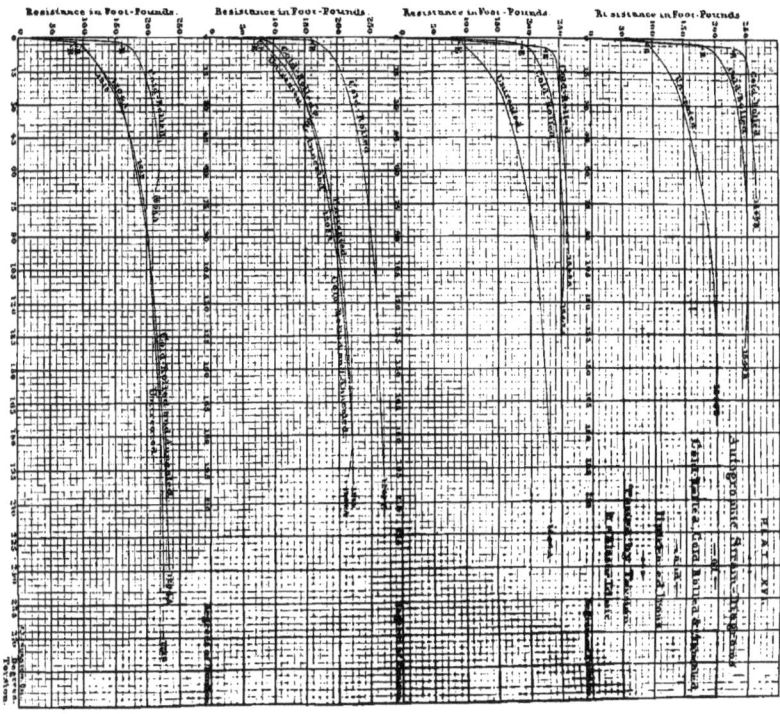

Resistance in Foot-Pounds.   Resistance in Foot-Pounds.   Resistance in Foot-Pounds.   Resistance in Foot-Pounds.

PLATE XVI.

PLATE XVII.

JONES & LAUGHLINS' IRON.

Cold-Rolled and Untreated. Before and After Test.

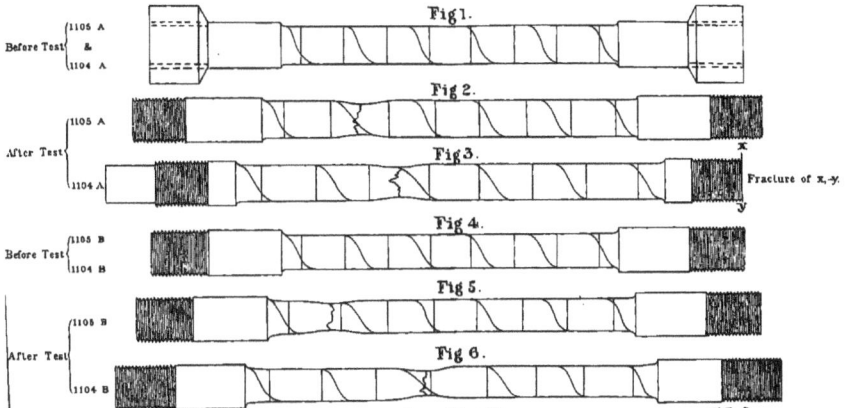

Fig 1.

Before Test { 1105 A & 1104 A

Fig 2.

After Test { 1105 A

Fig 3.

{ 1104 A

x

Fracture of x,-y.

y

Fig 4.

Before Test { 1105 B 1104 B

Fig 5.

After Test { 1105 B

Fig 6.

{ 1104 B

Scale. Half-Size.

Fig 7.

| 1105 A, 7/8 inch in diameter. | Cold-Rolled. See Plate 7. |
| 1104 A, 7/8 " " " | Untreated. " " 7. |
| 1105 B, 3/4 " " " | Cold-Rolled. " " 7. |
| 1104 B, 3/4 " " " | Untreated. " " 7. |

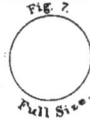

Full Size.